NGAMATEA

NGAMATEA
The Land and the People

HAZEL RISEBOROUGH

AUCKLAND UNIVERSITY PRESS

*For Lawrence and Winnie and Margaret with love,
for Jack who has kept the stories,
and for the succeeding generations whose history this is.*

First published 2006

Auckland University Press
University of Auckland
Private Bag 92019
Auckland
New Zealand
www.auckland.ac.nz/aup

© Hazel Riseborough, 2006

ISBN-10: 1 86940 369 X
ISBN-13: 978 1 86940 369 0

National Library of New Zealand Cataloguing-in-Publication Data
Riseborough, Hazel.
Ngamatea : the land and the people / Hazel Riseborough.
ISBN: 1-86940-369-X
Includes bibliographical references and index.
1. Farm life–New Zealand–Rangitikei District. 2. Hill farming–
New Zealand–Rangitikei District. 3. Ngamatea (N.Z.)–History.
I. Title.
636.3010993464–DC 22

Publication is assisted by the History Group, Ministry for Culture and Heritage.

This book is copyright. Apart from fair dealing for the purpose of private study,
research, criticism or review, as permitted under the Copyright Act, no part may
be reproduced by any process without prior permission of the publisher.

Cover photograph: Ren Apatu
Cover design: Athena Sommerfeld
Printed by Printlink Ltd, Wellington

Contents

List of maps	vi
Acknowledgements	vii
Introduction	1
One Owhaoko	7
Two Fernie Brothers and Roberts	29
Three Lawrence and Winnie	54
Four The Muster	88
Five Incidents and Accidents	127
Six Jack and Timahanga	159
Seven The Development Years	179
Eight Cooks, Cops and other Characters	211
Nine Margaret	242
Ten The Ngamatea Family	279
Notes	289
Bibliography	295
Index	297

List of Maps

1. Location, p. viii
2. Owhaoko Block, p. 6
3. Ngamatea 1930s–1960s, facing p. 88
4. Timahanga Station, p. 158
5. Ngamatea boundaries from mid-1970s, facing p. 178

Fernie–Roberts Family Tree, p. 31

Acknowledgements

I wish to thank all those who took part in the project for their time, their hospitality, their stories and their photos. Special thanks to the Roberts and Apatu families for their contribution towards expenses; to friends for encouragement and for hospitality as I moved between interviews; to Patrick Parsons, Rakei Taiaroa, Jinty Rorke and Ashley Gould; and to the staff at Archives New Zealand, the Alexander Turnbull Library, the Napier Museum and the Maori Land Court, Turangi.

Thanks too to Elizabeth Caffin of Auckland University Press and her skilful and friendly team, especially Katrina Duncan for typesetting; Louise Cotterall of the Geology Department, Auckland University, for map-making; and Anna Rogers of Christchurch for editing.

I am grateful to the Ministry for Culture and Heritage for a New Zealand History Research Trust Fund Award in History which enabled me to replace my steam-driven computer with one that was up-to-the-minute but full of drama and inexplicable tricks.

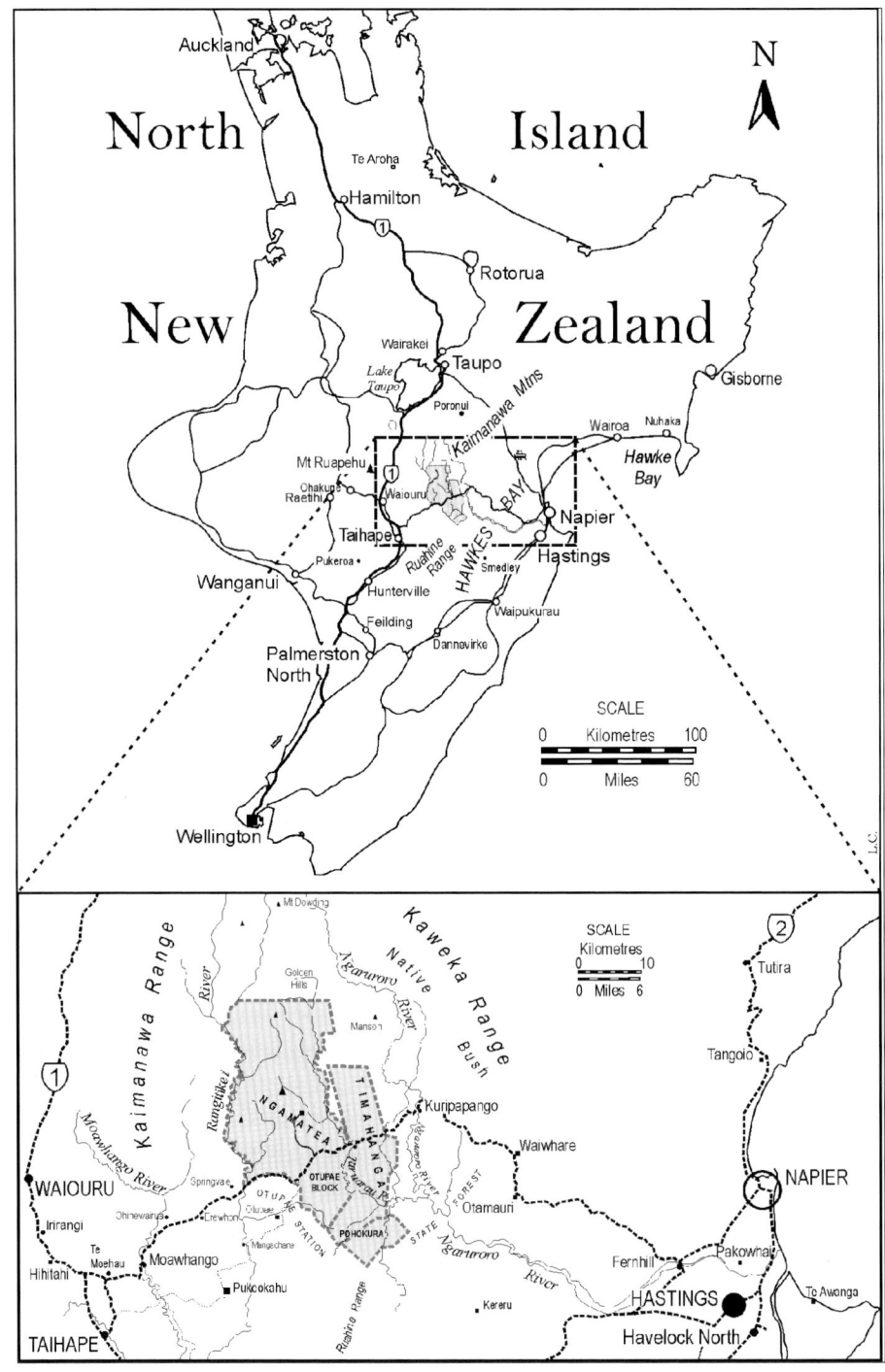

MAP 1: *Location*

Introduction

*You'll find anyone that's ever had anything to do
with Ngamatea likes to talk about it.*

Ngamatea station straddles the watershed of the central North Island. It is part of the Owhaoko block which stretches between the Kaimanawa and Kaweka ranges, and is bounded to the west and the east by the Rangitikei and Ngaruroro rivers, and to the south by the Napier–Taihape road. Fifty years ago Ngamatea sheep grazed perhaps 250,000 acres of tussock and scrub country with bush, mountain faces and river gorges, and only 5,000 acres of grassland. In 1972 the property was divided and the two lower-lying outstations, Timahanga and Pohokura, were then run as a separate station. Today Ngamatea covers about 70,000 acres, over 20,000 of which are highly productive grasslands. It is challenging country, high and remote and often snow-covered in winter. But the vastness somehow enfolds you and the tussock gets into your blood. They call it tussock fever and it infects almost everyone who goes there.

At least three major tribal groupings – from Hawke's Bay, Moawhango and Taupo – contested the ownership of the Owhaoko block when it went through the Native Land Court in the 1870s. That generated a parliamentary inquiry which occupied 150 pages in the parliamentary papers and led to legislation which resulted in a rehearing in 1888. The early lessees, from the 1870s to 1930, struggled to make the place pay. It overcame several of them, but by early 1931 Fernie Brothers and Roberts held Ngamatea and its two outstations and they alone survived the challenge. Drummond and Walter Fernie were graziers rather than farmers and the sheep they grazed were merinos and halfbreds. They bought stud merino rams in Australia, paying for them with handfuls of cash from an old sugar bag slung over one shoulder. Marvellously eccentric, they dressed like tramps, but were among the largest landowners in the country. There are endless stories about them, about their uncashed cheques and their pockets full of wool, about their meanness – and their largesse, and above all about their toughness. Walter would ride all night, from one of their properties to another to be on the job for the muster at dawn. He would arrive frozen to the saddle and his shepherds would have to pry him loose and lift him down, but he was ready for work.

Ngamatea was unlike any other North Island property in that a team of seven musterers and a packman-cook were employed for eight months of the year for the annual muster. They spent weeks at a time in back-country huts, with packhorses and teams of dogs, bringing the sheep in to the station for shearing, then returning them to the back country for the winter. It was the ambition of every young shepherd with a pack of dogs to do a season on Ngamatea. Some of them came back season after season. Some stayed through the tough winters between mustering seasons. Several general hands – rabbiters, gardeners – worked there year round for up to 20 years. Many of them remain connected to Ngamatea years after they put in time up there. They are part of the 'Ngamatea family', and this is their story, told in their idiom where possible. And it is particularly the story of Lawrence and Winnie Roberts who lived and farmed there for over 30 years, and of their children, Jack and Margaret, who grew up there and were moulded and honed by Ngamatea.

In 1959 when Margaret came to Massey Agricultural College to do the wool course I was instructing in the Wool Department. She soon suggested a visit to the station so we met her parents in Hastings and all set off in the Super Snipe, with Margaret at the wheel, to traverse the Hastings–Taihape road. 'I'd rather do it at night,' she announced, 'easier to see lights than dust.' I could not speak. Over Gentle Annie, up the Taruarau hill – I did not know which was worse. There was the Ngamatea gate, at last, but there were still another 7 miles to go. Margaret crawled under the car and wrestled the chains on. We started up the grass strip, ploughed out of the tussock as an alternative to the often impassable station road, slipping and sliding, swinging onto the tussock now and then, bump bump, to get a bit of traction, Margaret struggling with the wheel and hauling that heavy car back onto the Strip. Nobody spoke. On and on until suddenly Mrs Roberts, sitting beside me in the back, relaxed. 'It's all down hill from here,' she said quietly. 'We can slide the rest of the way.' From the top of that hill you can see Boyd Rock, way to the north above the upper Ngaruroro. The station lies spread before you, the great Ngamatea basin ringed by mountains, the tussock golden in the sun, a living entity, rippling and tossing in the breeze. From that moment the magic takes hold of you.

It is surprising that the story of Ngamatea has never been written. It rates a mention in books about back-country or high-country stations, and there are one or two memoirs written by musterers who spent a season or part of a season there, but it is not the easiest place to research. No one map is big enough to encompass the whole place: it lies between provinces, survey districts, land districts, Land Court jurisdictions, catchments. There are few family papers, apart from those of the first lessees, the Studholmes, who were there from the late 1870s; those collections are held by the family or deposited in the Alexander Turnbull Library. None of the other lessees left records. There are, however, a surprising number of official records relating to the Owhaoko block.

At the end of the First World War the government was looking for land for returning servicemen, and Tuwharetoa, burdened with survey liens and rabbit rates on remote unproductive land, gifted over 35,000 acres of the northern end of the Owhaoko block for soldier settlement. It was never able to be used for that purpose, and through the decades until the 1970s three government departments produced copious records of their intentions regarding the Gift Block: Lands and Survey were going to develop it; Forestry was going to plant it; Maori Affairs wanted the interests of its people considered. Ngamatea leased only 8,000 acres at the southern end of the Gift Block, but the Crown was also eyeing all the other Maori land leased by Ngamatea. Not until Ngamatea freeholded most of their leases and the Gift Block was returned in the 1970s to those who had gifted it, did the subject disappear from the records.

Although there are few station records, Winnie Roberts luckily left a collection of notes written sporadically in her later years – stories of high-country domestic life, of accidents, of some of the old-timers who worked up there and left their mark on the place. There are no records of seasonal activities, no diaries, just a few notebooks with stock numbers – and the station wages books from the 1930s through the 1960s, which are a treasure trove. It is the oral history that fills out the record.

The Fernies and the Roberts are particularly private people. On my second visit to the station I complained to Mr Roberts that I could not find them in the phone book. His reply was typical: 'We get enough phone calls without putting our name in the book'. He wasn't enamoured of the phone: his hearing had been damaged in the First World War, and he spoke with the characteristic Fernie stammer. He was the typical high-country man, not a great talker, a bit shy, reticent. He would ride all day without talking, at one with the space and the silence. A young musterer, new to the station, rode for miles with him one day without a word being spoken. The newcomer's eyes were swivelling from horizon to horizon, not a fence in sight, nor even a sheep, and when he asked, 'How big is this place, Mr Roberts?' he was told, 'I ddoon't know and no other bugguger does either.'

It was true. He knew how much freehold there was – 5,000 acres – and you could tot up the leasehold acres, but there were no fences, no boundaries except rivers, and if you pushed the sheep across the rivers they could spread as far as there was grazing for them. Ngamatea had grazing rights from the Crown over vast areas of the back country in return for paying rabbit rates, but everyone has their own idea of how big Ngamatea was, how far they rode from the southern end of the station, in the Pohokura valley, to Mangamingi hut at the northern end. They are all full of stories, and yet we had never collected them, never thought to get to work with the tape recorder – until Margaret, my friend Missy, died in February 2001, at only 60 years of age. We were all bereft.

It was at the gathering after her funeral that I recognised Lance Kennett from the photo on the cover of *Matt's Last Muster*, a book about a young shepherd who had worked on both Ngamatea and Timahanga and who had been killed soon after he left there in 1990. Lance had also worked on both properties and they had ridden out the back to trace the pack-tracks his late father had used in his time mustering on Ngamatea. I made my way through the huge crowd and introduced myself. Lance was disconsolate: 'They're all going . . . and the stories are going with them. Lawrence . . . Winnie . . . And now Margaret . . . We need a book.' I took a deep breath. 'We'd better find Jack', I said. We did, and in minutes we had hatched the plot and we started collecting the names of those who were still left, those who could give us the stories.

It took a while to get going. An early interview was with Gordon Mattson who had been on Ngamatea from 1934 to 1946. He had figured in many of Winnie Roberts' stories – and some of them we had on tape; she was interviewed in 1976 by the *Spectrum Documentary* team for National Radio. Gordon was 91 years old in 2003, and he still had great recall. Another 60 or so interviews followed, all around the North Island and even down in the South Island at Coldstream, Rangitata, with Joe and Sue Studholme. I am sorry about the many we missed, but we could not talk to them all. We tried to get a range of decades and occupations: owners and managers, friends and relations, shepherds and musterers, truck and tractor drivers, cooks and carpenters, a roadman, a drover, bookkeepers, fencers, horse breakers, mechanics . . .

The stories they told are of a world that has disappeared. The speed of life was governed by the speed of a horse; the station was a self-contained world – it had to be. Today the tarseal stretches further and further from the towns and shepherds and station hands have their own transport. No one arrives up there on horseback today, with a pack of dogs at his heels. People on stations everywhere will have met similar characters and know similar stories, though to townsfolk and today's generation the idiom may be confusing and the stories almost unreal. But they are of interest to all those fascinated by New Zealand as it was, by the romance of the back country, by the idea of musterers and packhorses, as well as helicopters and deer poachers, cops and trespassers. The more recent stories are of the transformation of poor pumice and tussock land into high-producing pasture – to the benefit of the country's economy; of the replacement of fine-woolled sheep by a high-lambing composite breed, of super-giant discs hitched to huge tractors, of truck-and-trailer units each hauling out 100 bales of wool or 40,000 lambs at 500 per load, of shepherds on quad bikes moving mobs of 5,000 ewes, of drive-on lime and superphosphate spreaders, of scanning ewes and artificial breeding of cattle; and always of deer stalking and dog trials.

By the early 1960s development had begun on neighbouring stations like Otupae and Mangaohane, lower lying country which the Studholmes had used as a lambing block and for wintering stock. At the end of the decade Ngamatea too was in development mode and the tussock was giving way to pasture; and the scrub on Timahanga was being cut again, some of it for the third time, and the blocks were being disced, fenced, supered and grazed to prevent further reversion to scrub.

Development on Ngamatea could not start until after the bulk of the station was freeholded in the mid-1950s and even then the place did not benefit, as many less isolated properties did, from three factors that revolutionised New Zealand farming around that time: giant discs, aerial top dressing and the Korean War wool boom. The roads could not accommodate heavy trucks; it was years before trucks with trailers could come over the Annie, and into the 1970s before there was a metal road that could take them the 7 miles in to the station; and it was a long and expensive haul to fly super in from Napier in DC3s. Then the Fernies were not innovative farmers so development did not begin until the first manager, Ray Birdsall, was appointed in the mid-1960s – and he had time to grow the first fat lambs ever sent off Ngamatea and get a cash flow going. The next great leap forward came in the late 1970s and early 1980s with land development loans, livestock retention incentives and supplementary minimum payments. Ngamatea's income no longer comes from fine-woolled sheep grazing tussock land; these days the station is 'a great big lamb factory', according to the present manager, Steve Kelleher. The back country has long gone, the days of packhorses and mustering teams are over. Only the stories remain. It has been a privilege to meet so many of the Ngamatea family, enjoy their hospitality and record their stories. I thank them all; I could not have done it without them.

Pukawa
August 2005

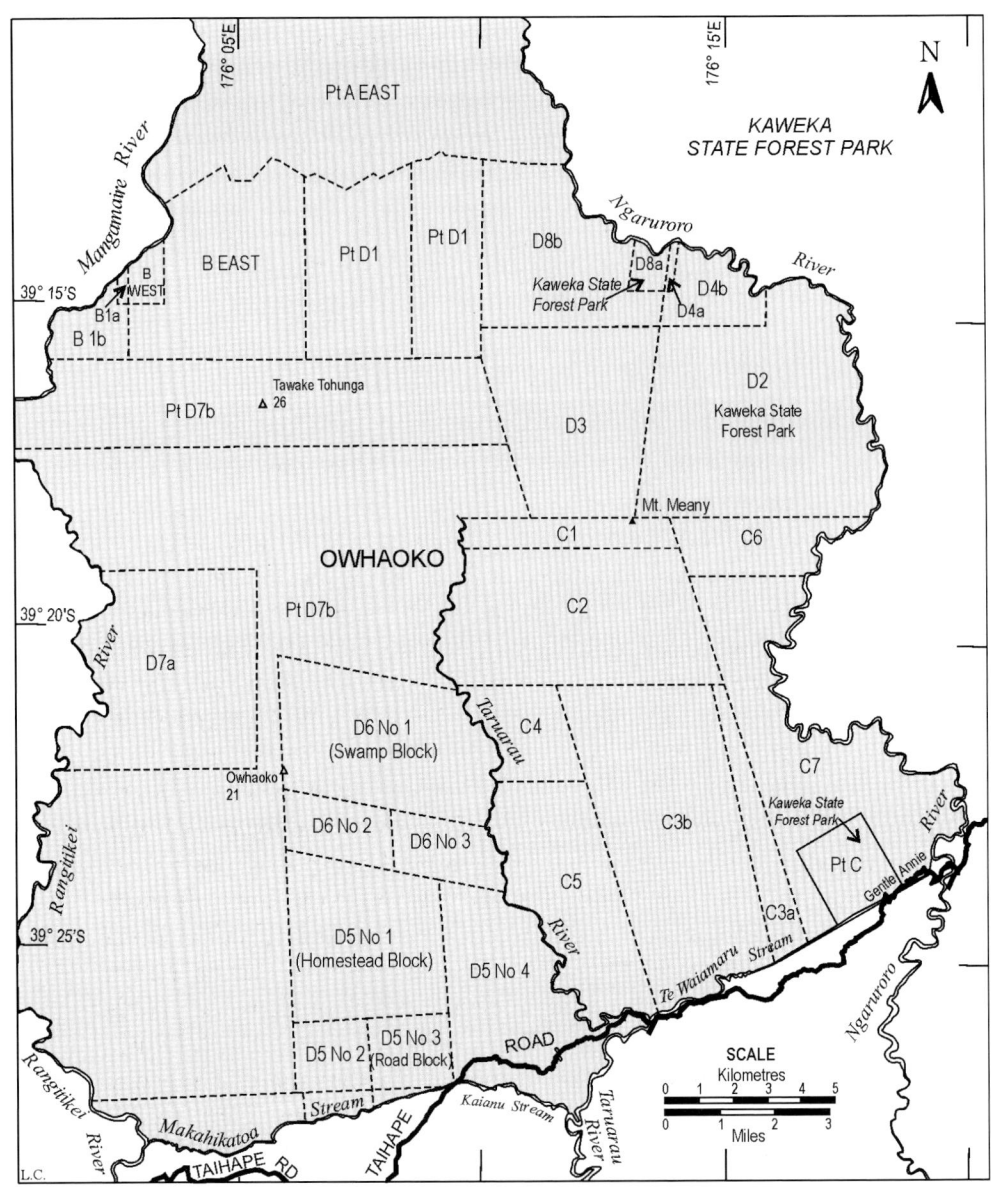

MAP 2: *Owhaoko Block*

Chapter One

Owhaoko

*From the nature of the soil and the climate the Owhaoko
block is unfit for residence.*

When Europeans first came to the Inland Patea to graze their sheep on the high tussock country there was little evidence of permanent occupation on the Owhaoko block. Maori from the Taupo region, Moawhango and Hawke's Bay had camps they used seasonally for fishing and hunting, but the high plateau, with its harsh winters, did not invite year-round settlement.

Maori history has it that the first person to visit and lay claim to the Inland Patea area was the great chief Tamateapokaiwhenua, father of Kahungunu, eponymous ancestor of the Hawke's Bay tribe.[1] At Whanganui a Roto, the inner harbour area at Napier, Tamatea gathered ngarara and koura – lizards and crayfish – which he carried in a calabash as he followed the Ngaruroro inland towards the northern Ruahine Range and crossed into the Inland Patea district. Along the way he deposited one of his little creatures here and there, thus bestowing place names throughout the region. Where the Ikawetea Stream runs into the Taruarau River, a tributary of the Ngaruroro, Tamatea left one of his pet lizards, named Pohokura. These mokai, or pets of Tamatea, acted as guardians of the district, keeping it safe for his descendants, Ngati Tama and Ngati Whiti.

Tamatea's son Tamakopiri and grandson Tuwhakaperei had come early to the area, and conquered the resident Ngati Hotu. They lived in peace for seven generations until another descendant, Whitikaupeka, led a heke, a migration, from Nukutaurua on the Mahia Peninsula to the headwaters of the Mohaka. There he established two pa, Rouiti and Rounui, on the Oamaru Stream, and fought with the neighbouring Kurapoto, a hapu of Ngati Tuwharetoa. The battle raged down the headwaters of the Ngaruroro and ended with Tuwharetoa taking over the land at the northern end of Owhaoko, while Whitikaupeka and his people moved south of the Golden Hills, across present-day Ngamatea, and into the Moawhango district to join up with their Ngati Tama relatives. In time they combined to drive Ngati Hotu from the district, and Ngati Tama moved to the western side of the Moawhango while Ngati Whiti settled on the eastern side.

The state of war between Ngati Whiti and Tuwharetoa continued, and when one of Whitikaupeka's descendants lost his life at the hands of Tuwharetoa, the hapu went to Rotoaira to exact utu. The battle went in their favour, and a high-ranking woman was captured and taken back to Moawhango where she was married into the hapu to cement a peace between the warring parties. That peace was never broken.

The earliest Europeans to visit the Inland Patea were missionaries. Richard Taylor from Putiki had come into the district from the west in 1845, and William Colenso had come across the Ruahines in 1847. On subsequent visits he followed the route via Kuripapango, the first European to do so. The sheep men were not far behind; by 1852 sheep were grazing in the Rotoaira area, and the Kuripapango route became the outlet from Hawke's Bay. Early in 1868 Azim and William Birch brought the first sheep, about 4,000 merinos, into the Inland Patea district using that route from Hawke's Bay. They had leased the 115,000-acre Kaimanawa–Oruamatua block between the Rangitikei and Moawhango rivers, to the west of the Owhaoko block. In time the block was divided between the two brothers, one keeping Oruamatua, the homestead block which included the present Ohinewairua station, and the other the Erewhon block. The third great leasehold block in the area, Mangaohane, was also later divided, the easternmost part, Otupae station, being cut off.

In the 1860s, too, a touch of gold fever hit the district. Goldminers from California and Australia were active in many parts of the country, and a few prospectors searched the rivers and ranges of the Inland Patea for traces of gold. They were apparently unsuccessful, but there were rumours of a find on the upper Ngaruroro, and Panoko Stream became Gold Creek. The Golden Hills, a little further south, took their name not from the precious metal, but from the glow of the tussock in the sun.

Owhaoko was first leased in 1875 to a man variously called Rainey, Maney and Marney, who may have been a storekeeper at Meeanee, but who described himself as a sheepfarmer of Omahu. The whole story of the ownership and leasing of Owhaoko is shrouded in this sort of confusion, but one or two things are certain. In 1875 Richard David Maney was elected to the Hawke's Bay Provincial Council for the district of Napier; and in 1876 he had an elegant lease drawn up with the purported owners of Owhaoko No. 1, a block of about 128,000 acres.[2] The boundaries are described and shown in a sketch map, and the lease is dated 27 November 1876 and signed by Maney and seven owners – Renata Kawepo, Noa Huke, Karaitiana te Rango, Ihakara te Raro, Horima Paerau, Te Hira and Retimana te Rango. The lease is for a 21-year term, at £500 per year, payable on the anniversary of the lease, 'the first of such yearly payments having been already made'. The lease was explained to the lessors by a court interpreter and the signatures witnessed, but it was never stamped. Pencil notes across the bottom in two or three hands record that it did not appear to be enrolled in the Native Land Court, as required by law. One problem was that the survey was incomplete; another was that the dates of the 'memorial of ownership issued by the Native Land Court' were not filled in. The first two Owhaoko blocks, only 28,782 acres, had gone through the court on 16 September 1875, on the application of the Hawke's Bay chiefs Renata Kawepo and Noa Huke, but the memorials of ownership were not issued until 2 December 1876. The main block of 134,650 acres was heard on 1 August 1876, and that memorial of ownership was issued on 30 October 1877, without Horima Paerau's name. So we are left with six more or less closely related owners for a block that obviously had dozens of owners from several tribes.[3]

The land Maney hoped he had leased was bounded by the Rangitikei, Mangamaire, Mangamingi and Ngaruroro rivers which in places cut down into deep gorges, but the bulk of the country ranged in altitude from 3,000 ft up to 5,665 feet at the top of Makorako, the highest peak in the Kaimanawa Range, and the fifth highest in the North Island. The block was basically a huge undulating tussock-covered basin around the 3,000- to 3,500-foot mark, draining to various streams or to the great Ngamatea Swamp and ringed by mountains. On the outer edges of the tussock country there were extensive areas of bush, largely red beech; the gullies and gorges were covered with scrub and low sub-alpine species; and the country ran up to steep rock slides, rocky outcrops and the mountaintops.

The early survey maps show a track running north and north-west for about 9 or 10 miles from the Inland Patea pack-track towards the Tikitiki bush, where the earliest station buildings and a woolwash were established. It is probable that sheep belonging to Renata Kawepo's people, and even to Maney, were already on the land before he negotiated his lease; local historian and runholder Tony Batley believed that Renata could have had a 'headquarters' there, and then maybe Maney had an outstation, but that the latter's 'headquarters' were nearer the site of the present Ngamatea station buildings, 6 or 7 miles in from the Napier–Taihape road. An archaeological survey carried out in 1981 by Tony Batley and Dr Ross McQueen, a botanist from Victoria University, showed 'a large rectangular ditch-and-bank enclosure ... divided internally by 4 discontinuous earth walls', which, with evidence of beech timber and even part of a mortised post, 'suggested internal division for stock'. The marked difference in the interior and exterior vegetation indicated higher fertility within the enclosure, and led to the conclusion that it was 'probably a sheepfold dating back to 1870s'.[4] The site is 'south of Ngamatea Swamp between two southern tributaries of Titau Stream', not far from the present station buildings; it could of course

John Studholme.
COLDSTREAM

have dated to any of the early farming activities on the Owhaoko block.

The first formal leases of the block were made to John and Michael Studholme on 5 October 1878 for a term of 21 years. The brothers, South Island runholders who already held leases of huge areas of the central North Island, leased about 150,000 acres in five blocks, from various combinations of five owners.[5] With a third brother, Paul, they had arrived at Lyttelton on the *Labuan* on 12 August 1851. In 1854 they bought the Waimate run in South Canterbury, which Michael took over and renamed Te Waimate; and in 1867 John and Michael bought Coldstream on the coast on the north bank of the Rangitata River. Paul had returned to Britain in 1858, leaving his brothers to farm in partnership for some years. John lived mostly on a small property on the outskirts of Christchurch, and travelled continually in the South and increasingly the North Island, visiting his business interests and numerous properties.[6] The partnership between the two brothers had been dissolved 'by mutual consent' on 31

The Owhaoko homestead, 1886.

July 1878, before the first leases were formalised, but it was only publicised six months later, perhaps to avoid delaying further an already fraught process of leasing.[7] John took over all the North Island leases as well as Coldstream and other South Island properties, and Michael retained Te Waimate, a more valuable property than all the others put together.

One of the brothers' other leasehold blocks was Murimutu, about 150,000 acres south of Ruapehu, running 40,000 merinos. It stretched across the Waiouru country, and adjoined Owhaoko, which was used as a staging point for the packhorses making the long journey from Murimutu to Napier. The Studholmes built homesteads on both stations, but they were occupied by managers. Owhaoko's first manager was E. H. Cameron, who came up 'on loan' from one of the South Island properties for a year or two at the start, and he was followed by C. H. Dowding, 'a cheery fellow'. The names of both are commemorated in the landscape: Mount Cameron (Te Iringa on the maps) and Cameron Spur on the eastern side of the block, and Mount Dowding (often Dowden) in the Kaimanawa Range. The last manager was R. T. Warren who, like Dowding, was a bachelor and had come from Coldstream. They were extraordinarily 'good and loyal' men.[8]

Warren, who was at Owhaoko for over 20 years from at least 1883,[9] also acted as John Studholme's agent. He was a prolific correspondent and kept his employer informed of everything that was going on, often writing on embossed 'OWHAOKO' paper, sometimes from the Hawke's Bay Club

The Owhaoko 20-stand woolshed, 1895; about 65,000 sheep were blade-shorn in the peak years.

in Napier; his handwriting is excruciating and almost indecipherable. He was kept busy over the years 'making payments to the Natives, dealing with Donnelly, meeting with solicitors', but he still kept a good eye on the property and reported the daily and seasonal round. He talked of mustering wethers from the back country: 'we would have started today, but for the rain'; in September 1888 'the Mt Michael wethers never looked better'; in November 1889 there were 'many lambs dropping now'. In 1891 his bullocks were 'clean knocked out and four of them dead' and he needed a bullock team to bring firewood in from the Tikitiki; the wild dogs were giving more trouble than usual: 'we killed five in the last fortnight'; and he was working on the house up at Ngamatia (as they spelled it): 'I must put the chimney in the gable and alter the sitting room. I have pulled the chimney down twice and it smokes worse each time.'[10] Warren lived in the Owhaoko (or Mangataramea) homestead, a substantial six-roomed house, with shingled roof and pressed metal ceilings, at Kaimoko bush – 'Boyd's bush' – near the eastern edge of the block. He built a smart picket fence and planted hundreds of 'English' trees, and good orchards; a few surviving trees were still bearing fruit 50 years later. In an 1886 photo taken with a group of the station workers outside the homestead Warren looks very smart in hat, jacket and tie, apparently his everyday dress.

The only Studholme resident on any of the North Island properties was Joseph, the youngest son of John Studholme, who in 1888 was settled on

Ruanui, an 11,000-acre Maori leasehold block north-west of Taihape. From this base he could supervise the running of the other stations. He gradually acquired the freehold of that and of neighbouring blocks and built up a property running about 50,000 sheep. Until the main trunk line was opened in 1908 Ruanui wool, too, was packed out via Owhaoko to Napier. Another son, Willie, studied law at Oxford, and became the family's in-house lawyer, basing himself at times at the Owhaoko homestead and keeping an eye on the legal side of all the leases – 'getting our titles through the Court', as he put it.

No sooner had the first Owhaoko blocks gone through the court in September 1875 than objections were lodged by claimants who had not had the chance to appear at the hearing. The whole sorry saga of subsequent hearings, petitions, falsified court records and failures to notify sittings is a judgement on the running of the Land Court in that era and on F. D. Fenton, the Chief Judge.[11]

The act of surveying the blocks was a further irritant that caused contention between the various claimant groups. Stephenson Percy Smith began a triangulation survey of the Inland Patea in 1871, but it was Renata Kawepo who ordered the first surveys of Mangaohane and Owhaoko, at the instigation of prospective lessees. There were surveys in 1875, 1876 and 1877, unopposed at least on the ground, but when one was begun in 1881 it ended in an armed confrontation between the Moawhango and Omahu people.

Tony Batley had the story from elders at Moawhango after the Second World War, when he 'made a point of visiting the old people in the village'. At the end of 1867 the Birch brothers wished to lease the Owhaoko block, but the local people, who deferred to Renata's mana, said, 'Renata may not be pleased – we will offer you the Kaimanawa-Oruamatua block'. When a survey was being carried out on the Mangaohane–Owhaoko boundary in 1881 Renata was not pleased, and he sent up an armed party from Omahu pa to put a stop to the survey. After detaining the surveyor and confiscating all his gear, which they sent back to Hawke's Bay, they built a small redoubt on the flat below Hiwiopapakai – Trig F on the maps – just north of the Otupae turn-off. When Tony Batley saw the redoubt in the 1950s it was a sort of rectangular earthwork that had not quite been completed when Renata's party went up to the Tikitiki bush on the Owhaoko block for Sunday karakia. The Moawhango people, armed with muskets, also went out there quietly while the service was in progress, swept up all the firearms which Renata's party had stacked in a heap at the edge of the clearing, surrounded the party from Omahu and packed them off back to Hawke's Bay.

In 'an interesting illustration of history and tradition' Tony found an entry in one of his grandfather's diaries, dated 21 February 1881, which confirmed this story. He was riding down to Napier: 'Met Patea natives coming back from Tikitiki after sending off Ngati Upokoiri'. Tony checked the date: 'that was a Monday morning so I have every reason to believe what Hakopa told me about the incident taking place during a prayer meeting on a Sunday'.

The Moawhango people also stated their claim to the land by other means. An 1884 map, drawn by Henry Ellison and used in the Land Court in 1888, showed a small kainga just inside the southern boundary of Owhaoko C3b, across the creek from the Inland Patea road a few hundred yards west of the entrance to Timahanga station, where the road begins to descend towards the Taruarau River. Tony Batley saw traces of houses there which he believed to be post-European and probably related to the opening up of the Inland Patea track: 'Some of the Moawhango people – probably just establishing their presence – stating that they were here, that they lived here. They had a post at Kuripapango too, a pou called Whitikaupeka.' It was all an assertion of their mana whenua, especially against the claims of the Omahu people.

The manipulated withdrawal of an application for a court rehearing in 1880 had resulted in the affirmation of the original decision to award the whole of the Owhaoko block to six owners, and this occasioned further protest, particularly from the Taupo people. But despite this the blocks were subdivided by the court in 1885, and this time they were awarded to just five owners, 'it being then admitted that Hira te Oke had no ancestral claim'. The orders for Crown grants would be issued only after 'proper surveys' were done.[12] The protests continued and in the following year 'a large mass of papers, with minutes of the Native Land Court, etc, concerning the Owhaoko and Kaimanawa Native Lands' was referred by the Minister of Native Affairs to Premier Robert Stout in his role as attorney-general. Stout managed to unravel the whole mess and wrote a 23-page memo highly critical of the operations of the court and its officers. He concluded that, to overturn what he described as a 'gross travesty of justice', legislation must be introduced ordering a rehearing of all the blocks. The government acted on his advice,[13] and the Owhaoko and Kaimanawa–Oruamatua Reinvestigation of Title Act, passed on 18 August 1886, returned the land to its pre-court status of native land and protected the extant leases – those of the Studholmes on Owhaoko and the Birchs on Kaimanawa–Oruamatua.

The land went back to the court in the middle of 1887,[14] and again a year later, and all the claimants finally had the chance to be heard. At every previous hearing Renata and a few of his close relatives were the only ones heard, and the land was always divided among them, with Renata being allotted by far the greatest share. In 1887, his claim through his tribe,

Ngati te Upokoiri, was rejected, the court ruling that Ngati te Upokoiri had 'neither ancestry, conquest, mana, nor occupation wherewith to support a claim to Owhaoko' and that if Renata was admitted to share in the lands, it could only be by the custom of aroha, which Ngati Whiti and Ngati Tama, whose claims were upheld, could exercise 'should they see fit to do so'.[15]

In 1888 the judges found that Renata *did* have a claim – by right of descent from Whitikaupeka, and by occupation through his grandfather Te Uamairangi, principal chief of Ngati te Upokoiri, who in his day 'possessed absolute mana' over the Owhaoko block. He 'had a house called Moatapuwaekura and also cultivated food' – presumably at the Tikitiki bush. This was 'the only evidence of any permanent occupation on the block' but the judges ruled that the stand Renata had made against the claims of Te Heuheu 'tended to retain this block in the hands of Ngati Whiti', that Renata's 'claim to share in the block was proven, and that his name was to be inserted in the list of owners.'[16] But it was too late for him to see his claim vindicated: he had died on 14 April 1888, a few weeks before the court began its sitting.

The court upheld the claim by conquest of Ngati Kurapoto and Ngati Maruahine, of Tuwharetoa, the original applicants for the rehearing, to the northern part of the block. Their 20,000 acres or so, with the boundary fixed 'at the watershed of the Taruarau', was to be known as Owhaoko A. Owhaoko B, a small western block on the lower Mangamaire estimated at 7,225 acres, was awarded to Ngati Tama. Owhaoko C, the eastern block between the Taruarau and the Ngaruroro rivers estimated at 36,125 acres, was awarded to Renata Kawepo, to Noa Huke and to Paramena te Naonao and Anaru te Wanikau and such of their co-claimants as could prove descent and occupation; and likewise to Airini Donnelly and her co-claimants. The largest share of the block, Owhaoko D, the central and southern portion estimated at 101,150 acres, was awarded to Ihakara te Raro, Retimana te Rango, Karaitiana te Rango and their co-claimants of Ngati Whiti and Ngati Whititama.

The total block, after resurveying, amounted to 163,432 acres – that is 'the greater portion of that tract of country lying between the Kaimanawa Mountains on the north and the Ruahine Range on the south . . . bounded on the east by the Ngaruroro River, and on the west by the Rangitikei. It is a country that, before flocks of sheep were placed there, was prized for its richness in native game, and for time immemorial has been a hunting field for the Natives for the kiore, kiwi, weka, titi, and the kakapo, and where eels still abound in the streams.'[17]

John Studholme had worked assiduously over the years to get the land passed through the court, and especially to get the leases renewed in the

Portrait of Renata Kawepo by Gottfried Lindauer.
COLDSTREAM

1890s. He had 'studied the Maori law as to leases, consulting Cornford, Rees, Bell and Moorhouse and reading through large numbers of Native Acts' so that he might know what restrictions there were on leasing, if any. Then he 'put the matter of obtaining a renewal of the lease of Owhaoko C and D into E Bell's hands and had a long talk to him on the best course to take'. He had travelled widely and had even written to the judges.[18] His papers are full of his travels and travails, but he had a particularly warm relationship with Renata Kawepo. They corresponded, and in 1885 he took his eldest son, Jack, to meet the great chief at his home at Omahu. In the Studholme homestead at Coldstream there is an imposing portrait of the chief, with cloak, huia feather and mere. It was painted by Gottfried Lindauer and 'legend has it' that Renata commissioned Lindauer to paint it 'as a measure of his esteem'.[19]

At first Owhaoko was worked in conjunction with Mangaohane, as the Studholme brothers had acquired the lease of 30,000 odd acres of that block, to the south of Owhaoko. It was a useful addition, being lower country used as a lambing block and for wintering stock, and for growing oats, but it became an area of contention between the Studholmes and the Donnellys and the source of protracted litigation in the Native Land Court. Airini Donnelly, married to George Prior Donnelly, was a great-niece of Renata Kawepo, who had brought her up.[20] In the 1888 court hearing she 'and her party', and Anaru te Wanikau and his, each claimed to have inherited Renata's right to Owhaoko.

John Studholme's great-grandson Joe remembered the family stories about Airini: 'She sat through all the court sessions, saying very little I think, but she was always there. One of the Studholmes did a sketch of her in court. She was a formidable presence – she had married an Irishman, and with the cleverness of the Irish and her ability and her position the combination was very tough.'[21] On 17 February 1894 John Studholme met Wi Broughton, Renata Kawepo's adopted son and one of the claimants to his interests, and Fraser, Broughton's agent, who 'promised to bring Block C before the Court . . . and to do all in their power to keep Mrs Donnelly out of the Homestead'. Then on 18 March he 'attended Broughton's funeral' and the next day 'saw Fraser in Napier and told him that though he was acting for Broughton in Owhaoko C, if he succeeded in getting Mrs Donnelly kept out of the Owhaoko homestead block I would make him a present to supplement anything Broughton gave him'.[22]

The 1880s were tough farming years. Although Owhaoko and Murimutu combined were shearing around 115,000 sheep, both merinos and halfbreds, Owhaoko was not paying its way and Murimutu was losing money. The finance needed for further development had to come from Coldstream, which had been 'allotted' to Jack in 1887: it was 'the only property making a fair profit at the time'. Another five properties, north and south, were also losing money. In 1889 John Studholme 'practically gave up control of all his runs and estates' to his sons and in 1890 they were able to freehold the bulk of the land they had leased at Mangaohane; in 1893–4 they shore 60,400 sheep on Owhaoko, and docked 15,000 lambs. Then in 1894 it seemed as though all the legal wrangling was behind them when they were successful in their attempts to renew all the Owhaoko leases – the D blocks for another 21-year term from 1894 and the A and C blocks for 14 years from 1899.[23]

John and Ellen Studholme had made many visits to England over the years, often with the family, and all three sons attended British universities. They went again in 1900, intending to return to New Zealand, but on 7 May 1903 John died there. Abruptly, in 1904 his sons relinquished all the renewed Owhaoko leases.[24] Willy had gone back to England to live, Jack was settled at Coldstream and Joe was at Ruanui, a profitable freehold property that stayed in the family until the 1920s. None had John Studholme's enthusiasm for the North Island properties so they let Mangaohane go too, selling 'the goodwill of the lease and the freehold . . . to this George Donnelly, and got over £30,000 net which was worth fighting for and made up for all our trouble and expense'.[25]

At this point in the history of the Owhaoko block the paper trail fades out. Luckily there are the accounts of one or two old-timers and some family reminiscences to help bridge the gap, but there are no other collections of family papers and few other official documents before the 1950s. Over the years the Crown had possessed itself of considerable areas of the back country, either by purchase or in satisfaction of survey dues, and it was more difficult now for prospective lessees to take up large areas of Owhaoko. Most of the blocks had been partitioned through the Native Land Court and required signatures from many Maori owners before any lease could be effected, but the Crown was anxious to see the country in production – and in European hands. It hoped to collect survey dues from rents, and early in the twentieth century rabbits began to infest the Inland Patea and became a further charge on the land.

Despite the obstacles, in 1906 Frederick Delannoy Luckie, then described as 'sheep-farmer of Ngamatia' but earlier as 'commission agent' of Hastings, applied to lease just under 60,000 acres of the Owhaoko front country, all Maori land. Such an application would be approved if the land was 'of poor quality or broken, or suitable for pastoral purposes only' in order 'to insure the taking up of such lands'. By the time the approval was granted a year later he had already begun to acquire his leases. Most of them were to run for 21 years, and one for 29 years, but by the middle of 1914 they had passed to Annie Ward Shaw, wife of Guy Landsdowne Shaw, and his sister, Maud Mary Landsdowne Shaw.[26] Guy already had at least part of Mangaohane No. 1,[27] so the Owhaoko leases may not have been taken out in his name because of the regulations limiting the amount of land any one person could hold.

Guy Shaw was a nephew of G. P. Donnelly, the son of his widowed sister Nan, and his uncle took an interest in his upbringing and education. When G. P. heard the British government intended to settle farmers on the Lesotho border he sent Guy off to South Africa with dogs and seed to make his fortune. He farmed there for about seven years, but locusts, hail and drought put paid to the fortune. However, in 1909 he married Annie Ward Mitchell, the daughter of a wealthy businessman, and Guy

The wedding of Guy Shaw and Annie Mitchell, May 1909, Ficksburg, South Africa.

and Annie moved back to New Zealand. Again G. P. helped his nephew to settle and he ended up farming part of Mangaohane and Ngamatea. Guy and Annie Shaw and their two small children, Beatrice and Basil, lived in the homestead at Ngamatea until the winter of 1918, when Guy drowned in the Woolwash dam. The inquest and coroner's report concluded that the cause of death was suicide. It was tough trying to farm that country: the lessees before and after him were driven off by lack of income, rabbits and debt. It had been a disastrous season and G. P. Donnelly had died the year before. There was even a suggestion that Guy Shaw may have been suffering from some illness, that while 'he normally weighed about fifteen or sixteen stone, at the time of his death he was a mere skeleton'.

According to John Spear, who knew him in 1915, Guy Shaw was 'one of nature's gentlemen and a really fine man to work for and with'. His widow Annie battled on by herself for two years until her parents came and persuaded her to go back home to South Africa. Many years later Beatrice,

Beatrice Shaw at Ngamatea homestead, c. 1916.

born in Napier in 1913, recorded her memories of her father – tall, blond and blue eyed – and her elegant mother in her 'perfectly tailored riding habit, boots, stock, and hat, enabling her to mount gracefully and sit comfortably on her side-saddle'. She recalled riding with her father: 'he had me in front of him on his saddle, and so we went rabbiting, coming home with rabbits hanging on a wire from Dad's saddle'. They lived happily and comfortably, the house full of fine furniture, etchings and oil paintings. 'Mum must have worked very hard for there was no help, yet she insisted that they dress every evening for dinner. After dinner every one retired to the lounge and gathered around the piano.' The young Beatrice had been told her father's death was accidental, and happily she never lost that belief.[28]

In 1920 all the leases passed to W. R. Gardner, 'flourmiller of Cust' in Canterbury. Gardner and his brother-in-law George Ruddenklau formed the Ngamatia Station Syndicate with A. B. Carmichael and Tudor N. Baker, which was dissolved in December 1923. Gardner and Charles and George Ruddenklau carried on the business. It must have been a rocky road. In September 1928 they created a company which took over some of the leases briefly until they passed to George Ruddenklau. The leases were mortgaged; things were not going well, and in 1934 Ngamatia Ltd was struck off the Companies Register.[29]

None of the Ruddenklau family lived on the station, although Charles's son, Karl, worked up there for about eighteen months in the 1920s. His son John remembered him talking about the Manson hut and Golden

A bullock wagon arriving at Ngamatea, late 1920s; Jock Watherston at back right.

Hills and that 'when Dad was shepherding up there they used to go to the Gretna in Taihape – that was their drinking hole'. Charles, Karl and the family accountant would drive up from Napier and back in a day, leaving at 5 a.m. and returning at 8.30 p.m., across that difficult Inland Patea road. In 1928 and 1929 there were many family discussions with the lawyer and accountant.[30]

Some of them may have involved negotiations with their manager, Jock Watherston, who lived in the Ngamatea homestead for five years with his wife and two small children. Jock's two sons, Bob and Ian Watherston, recall their parents talking about Ngamatea, but they 'only have photos and memories'. They remember talk of six-week musters, and that their mother cooked for the station staff, up to as many as 24 in the harvesting season, 'in absolute privation'.

Jock Watherston had taken over about 1924, when the place was very run down and the sheep in poor condition. He collected his own staff around him, including his two brothers, Donald and Jim, and in a few years they had the place in good shape, despite the rabbit infestation of the 1920s. Central Otago men, they had had plenty of experience with rabbits, and with fire: they set light to the tussock and it burned for about six weeks. Bob and Ian Watherston believed that the Ruddenklaus wanted to quit the property and that their father 'wanted to buy the place, but for what it was worth, not for government valuation'. In 1928 Jock Watherston applied to lease thousands of acres of the back country – 'unoccupied Native land'

Ike Robin and Jock Watherston, Ngamatea, late 1920s.

– rent free for the first five years, then £50 per year, as well as adjacent Crown lands. The officials noted that the land was rabbit infested, that the Crown spent £400 a year there on rabbit control, and that if it was possible to save that sum annually by leasing the land 'to some reliable person', then the application should be given favourable consideration. But as Watherston was unable to come to an agreement with either the Ruddenklaus or the Crown he and his family left the property in 1929 and another manager took over for a short time.[31]

At the beginning of 1931 the Ruddenklaus were carting wool and skins down to Hawke's Bay, and a week later got approval to lease Owhaoko D5 No. 4 – a 5,500-acre block bounded by the Taruarau Hill and Taruarau River.[32] On 12 February 1931, just days after the Hawke's Bay earthquake, Charles Ruddenklau 'went up for a few days'. It was his last recorded visit.

Times were getting tough, rabbits were at plague proportions, and there were limitations now on what the Crown could do about 'utilizing the land and keeping the rabbits under control'. Over the years the owners of various blocks, which they had 'vainly been trying to lease', had offered the land to the Crown, but at 'exaggerated' prices: in 1910 they were

Wild sheep; Jock Watherston at left.

asking 25/-, even £2 an acre, and the Crown, intent on getting the price down to 5/- told them 'they would have to accept responsibility of keeping rabbit numbers down and paying country rates'. When some owners were prepared to sell but those 'owning the subdivision where Studholme's old homestead stands objected to sell at the same price as the others', the Crown said if it could not get the whole block, it would take none. The owners were then reminded of their obligations under the Noxious Weeds Act and Rabbit Act, and told if they would neither sell nor use the lands themselves they would be 'made to comply with the onerous obligations that attach to lands of this nature'.[33] In those years huge areas of the back country were both unleaseable and unusable, except as a traditional seasonal hunting and fishing resource, yet the Crown would take no further action beyond piling up various taxes and rates as a charge against the land.

Then in 1917 the owners of a 4,000-acre block in the headwaters of the Taruarau found a very useful solution in recently enacted legislation: they gifted the block to the Crown for settlement of returned Maori soldiers. The Crown was delighted at this generous gesture on the part of its loyal subjects and the Governor proclaimed the land to be vested in His Majesty the King and to be Crown land.[34] This gift was 'treated largely as a tribal matter, the principal people involved being Te Heuheu Tukino, Kingi Topia, Te Hiraka Pine and Ngahuia',[35] and it was hoped this would be an example for other tribes to follow. They did: the Crown became the owner of 35,583 acres of 'tussocky, wind swept, poor quality land, ranging from

3,000 to 5,000 ft between the upper Ngaruroro and Mangamaire Stream', with no legal access, and about '9 miles from the Station which is 5½ miles from the Napier–Taihape road'. Although gifted specifically for soldier settlement, it was 'later found unsuitable for this purpose because of steepness and poor quality, and consequently was never used for settlement'.[36]

But under the trust set up at the time it could not be used for anything else either and it certainly could not be leased to Pakeha farmers. Over the years the Crown had not prevented lessees of the front country from grazing the back country. It was hard to stop them: there were no fences, and the lessees got 'free grazing' in return for paying the rabbit rates. In 1930, apparently without consulting the former owners, the Crown removed the trust by special legislation, leaving the area to be administered as Crown land. All revenues were to be paid into a special fund to assist Maori veterans or their successors.[37] This move did not help them lease the land, but it did provide a fruitful source of conflict between several government departments for years to come: Lands and Survey were going to develop the land; Forestry was going to plant it; Maori Affairs wanted the interests of its people considered.

From the early days Owhaoko station had had a neighbour in the 26,000-acre Timahanga block to the south-east. John James McNeil Boyd, a licensed interpreter, had explored the block with Anaru te Wanikau in 1885, and the following year, in partnership with him, he put sheep on the block and established a homestead there at Ruhanui (known today as Jimmy Leonard's). In a few years he had built three whare, erected sheepyards, 4 or 5 miles of fencing and 'some scrub fencing in places', and was running about 10,000 sheep. It was obvious there was no formal lease; Anaru, a close relation of Renata Kawepo,[38] was one of the owners of the land and in 1894 Boyd told the Land Court that no one had 'questioned Anaru's right to be there since 1886'.[39]

It appears that John Boyd moved from Timahanga into the abandoned Studholme homestead some years before he was killed in a fall from his horse in 1910, and that block, which had been known as 'Old Owhaoko' or Mangataramea, became known as Boyd's. There is no evidence that John Boyd took over the lease relinquished by the Studholmes. After his death his brother George took over his property and his 'land interests' in the Inland Patea. An early account of the area written many years later by John Spear, who lived up there from 1910, recorded that Gordon, one of George Boyd's five sons, managed the block in 1910 and 1911. Another son, Arthur, was squatting on the 21,000-acre Te Koau block, south of Timahanga, at the bottom of the Pohokura Valley, at the foot of the northern Ruahine Range.[40]

Between 1911 and 1915 the Crown was busy purchasing the Timahanga block, bit by bit, from those whose ownership had been decided by the Land Court in 1895. It became Pastoral Run 23 and Jimmy Leonard, among others, leased part of it from the Crown for the next fifteen or so years. Jimmy was another who left fruit trees behind: he had two orchards, one on a little terrace down near Jimmy's Creek, and another on the old track, the line of the present road. Some of those trees survived for years: 'We used to get quinces and cherries, I think – but the possums decimated the lot and killed the trees as well'.[41] About 1912 Arthur Boyd moved into the Studholme homestead and claimed the house and any station buildings that remained as his own. He married Jessie Kirkman from Hastings, and their son, Johnny, was born in 1919. But the marriage did not last; when Johnny was seven he and his mother left and moved down to Hastings. On two visits back to Timahanga about five years before he died in November 1995, he was taken to see his old home and was recorded on tape by Jack Roberts and, on the later visit, on video by Bob Ralph. The emotion in his voice is palpable. Driving up from Timahanga towards the old house he talked of the outbuildings, the car shed which 'had no door on it and the wild pigs used to come in and the old boars used to get under the car and dig a hole because sometimes it was warmer from the car engine, and they'd spend the night in there'. He just remembered the last of the bullocks 'but they were released and they were wild then. We'd picked up old shoes and we couldn't figure out what the shoes were, but they must have shod the bullocks for bush work. They had a big cleat in the front so they could get traction.'

When he got to the house which was then derelict and barely standing, he was almost speechless – 'she's deteriorated since I came up here last' – but he described the house as he walked through it: 'There was a chimney up there, Mum had a big stove inside there, and there was a window just there. My dad used to sit at the end of a long table that went along there, with his rifle and the pigs would amble along here and he'd just put the rifle out of the door and – pork! And he'd do practically the same thing off the front verandah with deer, if you wanted them. But there were so many and they were so accessible, you didn't shoot a lot.' He pointed out the bedrooms, and the bathroom in between – but the galvanised tin bath had been damaged before their time and his dad never fixed it, never put water on. 'The old man dug a well up there, it was beautiful water and we used to carry it in 4-gallon tins down to the house. He never put a pipe or a flume in or anything. And the toilet was way down by the bloody creek . . .

I used to be scared stiff when I was a kid, going down to that blasted thing. The moreporks would be singing out.' They did have a tank to catch water from the roof. 'My mum was getting water from it one day and she heard some people talking up the top of the hill here. Some young lads about eighteen came over the top and one of them put a shot between her legs. If the old man had been home he would have shot them – bloody vile temper, he wouldn't have thought twice.'

The kitchen and the dining room evoked special memories for him – the big table that went nearly all along one wall, the shelves where his mother put the bread to rise. 'She used to bake it in big loaves and they would pack a lot of them to the guys out the back. They called them cart wheels, and they'd last – they'd be fresh in five days' time – beautiful.' Then they came to the sitting room, where they would 'sit in the evenings with a big fire going. People used to visit quite often for those days – jokers used to come down from Mangaohane and Ngamatea, Otupae, and all over the place. Mum was the only woman for miles around then and she was a good cook, and she used to put on a great meal for them. There'd be snow all around here and it would be warm as toast.' And they had a radio, the only one in the district, a huge thing with about eight dials that nearly took up the whole wall. 'He could get the boxing and wrestling in America, clear as anything. We thought everybody could do it but they couldn't, and old Hughie Lyall, he was up at Otupae, used to come over and listen to the radio.' That was about a 40-mile round trip, over an excruciating road, but the Boyd homestead was the social centre of the district.

His father was an awkward character though and there was no warmth in his son's voice when he spoke about him. 'My old man was here and there was quite a bit of stock on the place – not much by your standards – but there was a bit. He didn't farm it at all – he did bugger all, really.' He raised a few pigs, but the wild boars would get in and breed with them. He milked a few cows, and his small son 'used to have to go up there to get the bloody cows of a morning, cunning old buggers they were', and then had to wrestle the cream cans up to the gate to catch the transport going by to take it down to Hawke's Bay. He just vaguely remembered mobs of sheep too, when he was very small, and one of them jumped the fence: 'people don't believe it today. Those halfbred merinos, they're bloody wild. They'd attack the men, and they'd even attack the horses, and some of them could jump fences like a deer.'

Spear claimed Arthur Boyd held the lease of the Golden Hills and wanted Spear 'to take the block off his hands'. Apparently he did graze sheep out there – after all, his name is recorded in the landscape – but the most the lease amounted to was a small annual payment and an agreement to control the rabbits. There was already a rabbit-proof fence at the back

The Owhaoko ('Boyd') homestead, January 2004.

of the Ngamatea Swamp, right across the property from the Taruarau to the Rangitikei; there are still traces of it today.

Johnny Boyd's 'townie' uncles, two of his mother's brothers, were good value and they would come up and do odd jobs and 'go out and have a shot'. But one managed 'to shoot the other in the backside with a shotgun – he only just got him, thank goodness, otherwise he would have blown his bloody arse off'. One of the uncles was a butcher, 'a helluva nice bloke, and he used to come up to stay. He was good company.' One day 'the old man' brought a bull calf in from the back and tied it up to a tree near the house. His Uncle Geoff said he would feed it, but it was 'a fair-sized thing and God, it was wild. Well, the calf took to him he didn't know where to go, so he went round and round the tree until he got the calf wound up and then he collapsed on the ground. He'd already dropped the bloody milk. Oh, God, I'll never forget that.' Incidents like that were amusing memories, but usually he said very little about the early days and his 'terrible childhood'.[42]

In 1916 Arthur Boyd was in negotiation with the owners of Owhaoko C3 to buy the 'homestead' block. Some of the owners agreed to sell, but several objected to selling to Boyd. When they all agreed to sell to him in 1918 'there was no word from Boyd'. In 1921 when the court registrar contacted his solicitor to ask whether the sale was proceeding he was told they had not heard from Boyd, and that there was 'no prospect whatever of his being able to carry out the purchase of this land'.

Arthur Boyd continued to squat there, growing ever more cantankerous, and in February 1933 one Hauparoa Hamilton, who had interests in the block, went up with a couple of friends for a weekend's shooting. They were careful to demonstrate that their guns were not loaded, but Boyd terrified them by wandering around with a .303 loaded and cocked. Hamilton wrote to the Land Court to complain that Arthur Boyd had been occupying the property for 20 years, paying no rent or rates, and that the place was being destroyed by neglect: 'Boyd is threatening anyone who comes in with loaded guns . . . we think he is mad, not criminal'. He wrote again a couple of weeks later to say he had inherited his grandmother's and mother's interests, and 'I can't occupy it while Arthur Boyd is at large with a 303 . . . please assist me to remove him'. The court replied that Arthur Boyd appeared to have no legal right to occupy the land, and that the beneficial owners were the only persons who could take steps to remove him. 'If he is menacing people as you suggest then report the matter to police . . . . You don't appear to have the right to go on the land yourself as succession orders have not been made.'[43] Hamilton must have sorted out the question of succession because a few months later he wrote to the Aotea Land Board to say he had given Boyd one month's notice 'to shift all his belongings off the land', but Boyd would not take any notice and still thought he owned the land.[44] Boyd got the message eventually, and by the end of 1933 he is said to have moved out and driven his sheep down to Stortford Lodge for sale; his father's death that year may have influenced his decision.

The 'Boyd' or Studholme property was never lived in again. Hunters used it for some years, but gradually it deteriorated. The iron that had been put over the original shingle roof was filched during the Second World War and water got into the cob chimneys and broke them down, and the surrounding trees added to the damage. It had been well built, though and it stood for many years, until finally the roof collapsed and the old house, with all its history, sank back into the ground.

Chapter Two

Fernie Brothers and Roberts

What do you think Walter Fernie bought himself for his ninetieth birthday? A new saddle!

After George Donnelly died in 1917 Mangaohane station was broken up. The Otupae block of 16,000 odd acres was freeholded by Donnelly's manager, Tom Morrin, and eventually sold to H. B. Williams. In 1920 the rest of the property was sold to the partnership of Chambers and McLean;[1] the sale included the Pohokura block – an outstation of Mangaohane – and the Otupae Range, and these two blocks of just on 14,000 acres later passed into the hands of William Drummond Fernie.[2] In about 1932 he also acquired the freehold of Te Koau, a block of about 14,000 acres south-east of Pohokura, across the Ikawetea Stream.

The Fernie brothers, Christopher, Drummond, David and Walter, already held large properties across the lower North Island. The family originated in Scotland, at Fernie Castle near Cupar, Fifeshire, where John Fernie was born in 1828. He came to New Zealand possibly in 1854, then returned to England for two years where he married his cousin Annie (Agnes) Norwood. He brought her back to New Zealand with their first child, Christopher, who was born in 1859. John, a civil engineer, later worked on the first Whanganui River bridge, but they settled on a 100-acre block which he had bought at Kaitoke near Wanganui, and which they called Churchill. They had another eight children, all born and brought up at Churchill: John, William Drummond, David, Jane Norwood, Annie Norwood, Katherine Elizabeth, Maud Mary and Walter Edward, born in 1877. John died aged twelve and was buried at Churchill. Drummond, born in 1861, was 'the general': he carried his little brother Walter on the saddle in front of him. There are plenty of stories about Drummond, and more about Walter, but even these are 'a bit hazy.... They just wouldn't talk. Walter would never talk – he would just grunt. What we've got is what we've picked up bit by bit.'[3] These are the strong, silent Fernies; not talking much is a family characteristic.

John Fernie set to work to clear his section of bush, and gradually acquired neighbouring sections until he ended up with about 500 acres on which he built a house, a woolshed and other farm buildings. Then he expanded from Churchill to Rusthall, a 640-acre block nearby on No. 2 Line. It is thought this property may have been given to his eldest son and that he finished up there, but very little is recorded about Christopher. As the other boys grew up they needed to expand further, and in 1881 they began leasing property up the Parapara, then in 1900 they bought Otamoa, 20 or so miles up from Wanganui, and later Pukeroa in the Turakina Valley north of Hunterville. By that time they were winning prizes at the local A and P shows for merino ewes, and merino, English Leicester and Corriedale rams.

Pukeroa was Drummond's property, Otamoa Walter's. Their great-nephews believe they had an agreement that if any of them got married 'the others would kick him out'. Drummond never married, but he came close to it, as John Roberts explained.

> He built this big house at Pukeroa, at the back of Hunterville. He was going to marry – he was trotting this girl – and one night he decided to go over and visit his lady friend. He tied his horse up at the gate, went up to the back door, hand up, just about to knock. The back door was open and he heard his lady friend telling her mother what she was going to do with his money. He never knocked, he just turned around, got on his horse, and that was the last he saw of her.

Walter married twice, but he had no children. His first wife, Helen Forsyth, was a farmer's daughter from Raetihi, 'a lovely looking girl', who died of cancer of the spine in 1948. Later he married his English cousin, Eleanor Hallas. They had met during the First World War when Walter stayed at her home while on leave, and they saw each other again over the years when he made other trips to England.[4]

David, the first of the brothers to marry, bought Chesterhope at Pakowhai, and Moeangiangi station between Napier and Wairoa. For a while he was in partnership with William Richmond, but the partnership was dissolved in October 1931 and David carried on alone.[5] He married Alice Arnett from Tangoio, and their only child, Joan, still has both properties. She never married; as they say, they are 'not a marrying family'. There were only three children in the next generation: Joan Fernie and Lawrence and Jack Roberts. Only two of the Fernie girls married: Maud Mary married Tom Wright in 1926 when she was 56, and Annie married Joseph Roberts on 30 November 1896. His parents were William Roberts and Ellen Abraham; they had married in Ireland, when he was nineteen and she seventeen. William could not write: he signed his marriage certificate with a cross. They had five children and William was killed in 1876 in

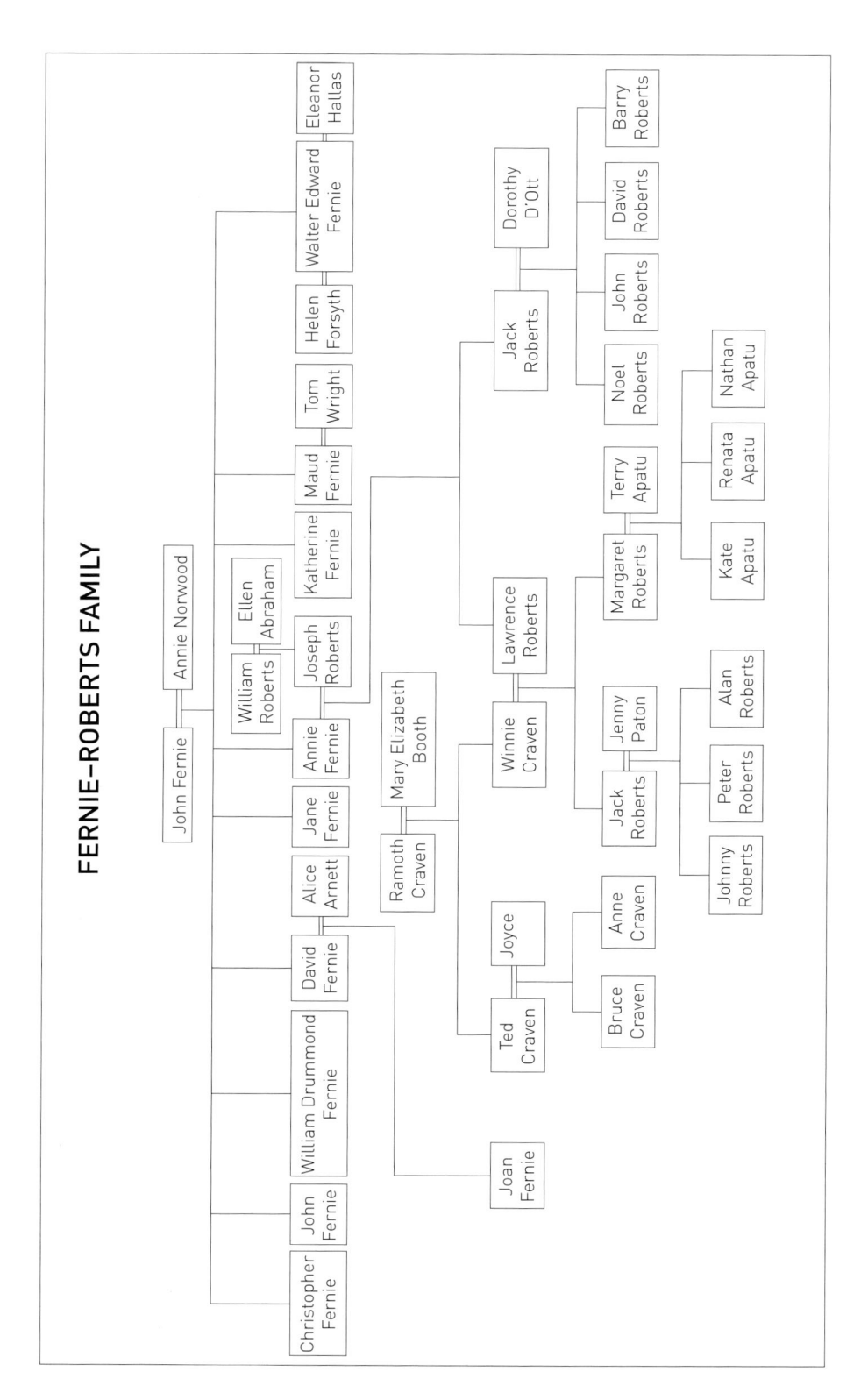

FERNIE–ROBERTS FAMILY

John Fernie — Annie Norwood

Children:
- Christopher Fernie
- John Fernie
- William Drummond Fernie
- David Fernie — Alice Arnett
- Jane Fernie
- Annie Fernie — Joseph Roberts
- William Roberts — Ellen Abraham
- Katherine Fernie
- Maud Fernie — Tom Wright
- Helen Forsyth — Walter Edward Fernie — Eleanor Hallas

David Fernie & Alice Arnett:
- Joan Fernie

Annie Fernie & Joseph Roberts → Ramoth Craven — Mary Elizabeth Booth
- Ted Craven — Joyce
  - Bruce Craven
  - Anne Craven
- Winnie Craven — Lawrence Roberts
  - Jack Roberts — Jenny Paton
    - Johnny Roberts
    - Peter Roberts
    - Alan Roberts
  - Margaret Roberts — Terry Apatu
    - Kate Apatu
    - Renata Apatu
    - Nathan Apatu
  - Jack Roberts — Dorothy D'Ott
    - Noel Roberts
    - John Roberts
    - David Roberts
    - Barry Roberts

FERNIE BROTHERS AND ROBERTS

a rock-fall at a Wanganui quarry; he was 36. Joseph cannot have had an easy upbringing. He worked on farms in the Wanganui district and, after his marriage to Annie, was employed for a time by the Fernies, but he died on 7 March 1901, aged 35, so another young widow was left with small children to bring up – Lawrence Harper Roberts, born on 25 December 1897, and John Fernie Roberts (Jack), born on 23 January 1899.[6]

Annie took the two children to Chesterhope and kept house for her brother David, but when he married she and the children moved out, although the Fernie men were always involved in the boys' upbringing. Annie remained in the district and Lawrence and Jack went to Pakowhai School. It was there they met Ted and Winnie Craven. Winnie was born in Bradford on 5 October 1899; Ted was a couple of years older. The Cravens, Ramoth and Mary Elizabeth and the two children, had come to New Zealand from England when Winnie was three. For a start they had the Pakowhai store and the coaching service between Napier and Pakowhai or Napier and Hastings and they may have had an orchard there for a while. Later they had a dairy farm at Irongate. Lawrence used to sit behind Winnie in school 'and dunk her plait in the ink well', and from their school days on the two families kept in touch.[7]

Lawrence always carried a torch for Winnie, but she was too involved with life, too busy with horses and her other interests to get serious over any boy. According to his son Jack, when Lawrence was 20 he enlisted in the Army as a member of the New Zealand Rifle Brigade.

> Uncle Walter was at the First World War too. I don't think they went together, but they both served there. Dad was a gunner/driver on the 18-pounder guns. He said very little about his war experience. The only time he told me anything was one day he said that the German guns had got their range and landed a direct hit on one of their gun pits. The guns had to be evacuated as soon as possible and when they took the guns out they had to take them over the bodies of the soldiers and gunners who had either been killed or wounded.

It may have been then that Lawrence's hearing was damaged. His discharge certificate issued in August 1919 described him on enlistment as 5 feet 7 inches tall, fair complexion, blue eyes, brown hair, a shepherd by occupation, and showed he served abroad for two years and two days. Lawrence carried a photo of Winnie all the time he was away – and there in her photo album, in the centre of the first page, is a smiling friendly photo of Lawrence in uniform by the Napier sea-wall.

After Lawrence's war service the 'go-ahead' Drummond settled him as manager and partner on Mangatapiri, one of his Hawke's Bay properties near Elsthorpe. Lawrence's younger brother Jack had married Dorothy

D'Ott and they lived with their two boys Noel and Johnny, at Otamoa, then moved to Haumoana, where her parents lived. Drummond had the freehold of Pohokura and Te Koau, and he took over the lease of Timahanga from Jimmy Leonard; it was a Crown lease of part of Pastoral Run 23. He made Jack manager and partner of these Inland Patea properties.

Timahanga was a 7,891-acre block. A long, narrow, sheltered valley, bounded to the north by Kaimoko bush (part of Owhaoko C3b), it is cut in two by Jimmy's Creek (Timahanga Stream) and separated from Pohokura by the Taruarau River. Pohokura, a rich block at the bottom of the valley, has a beautiful area of bush, the Paramahao, and is separated by the Ikawetea Stream from the Te Koau block which runs up onto the northern Ruahine Range (see Map 4, p. 158). The blocks are bounded by the Comet and Ruahine ranges to the east and south-east, the Sparrowhawk Range to the north-west, and the Otupae Range to the west. The altitude varies from 2,350 feet at the homestead in the north-east corner of the block, down to 1,350 feet at the river. Access to the valley was very difficult. It is 10 miles from Timahanga homestead to Pohokura and there was just a pack-track through the scrub and then Jimmy's Creek, which had to be forded 30 or more times; it took several hours to get from Timahanga down to Pohokura with packhorses. The station headquarters, though, are right beside the Taihape–Napier road, almost literally just over the fence from the old Studholme homestead where Arthur Boyd was still in residence.

In 1929 Jack Roberts left his family down in Hawke's Bay and moved up to Timahanga to supervise scrub cutting and begin to bring the block into production. At the end of 1930 Seth Barnes, who had first worked for the Fernies at Churchill, started up there cutting scrub and almost immediately became a general hand, then the overseer at Timahanga until 1934. Three of Jack's brothers-in-law – Hughie, George and Les D'Ott – were all up there at one time or another; Les was a good horseman who had broken horses on Mangaohane. Initially the men had to drive the sheep from Timahanga and Pohokura over the range to be shorn at Otupae station but then they constructed their own station buildings. All they had for the first couple of years was a small corrugated iron cookhouse, but Drummond had big plans. The mill at Taihape could not supply the 60,000 feet of timber he ordered and had to share the contract with another mill. Walter Bates, known to all as Chips, became the station carpenter in April 1931 and he and his assistant got to work building an eight-stand woolshed, a cookhouse, shearers' quarters and stables; some of those buildings are still standing to this day. A South African War veteran and a meticulous worker, Chips put in years of service for the Fernies. The wages book shows that Arthur Boyd found work there too for four or five months in 1931–2. A scrub-cutting gang of 'Hindus' (actually Sikhs) from India was soon at work at Jimmy's. They were a self-contained crew and lived at the 'Indian Camp' down there. The boss, Jack Roberts, camped too. There would be no house at Timahanga for another 35 years.

When the Hawke's Bay earthquake struck on 3 February 1931, Dorothy Roberts and her two boys were living with her people at Haumoana. Jack and Seth Barnes and one or two others from Timahanga piled into Jack's three-seater Chrysler and chanced their luck on the road. Somehow they got through, damaged bridges and all, and Jack found his family safe. He put up his tent outside the house at Haumoana and they slept in that – 'cats and dogs and everybody in there'. As soon as he could, he moved his family up to Timahanga. The cookhouse, now lined and spruced up minimally, became their home and Dorothy's kitchen, but they slept under canvas inside a shelter with a tin roof and stakes and bundles of scrub round the sides. 'I woke up in the tent and wondered why it was so light,' Noel remembered, 'and there was six inches of snow everywhere. That was my earliest recollection. I was five years old.'

By early 1931 the neighbours at Ngamatea had had enough of rabbits and depression and intended to quit the property. It was said that the government approached Drummond Fernie to see if he would be interested in taking the place over, and that Drummond agreed provided he paid no tax on the property for ten years. The story could well be true; the government certainly wanted the country farmed and producing and the Fernies were a pretty good bet to make a go of it. Drummond negotiated with the Ruddenklaus, took over the existing leases and bought the 5,000-acre homestead block, the only freehold on the property.[8]

Jack and Dorothy Roberts and their family spent a Christmas at Timahanga then moved up to Ngamatea early in 1932. Noel's next recollection was of the move. 'I remember when we turned off the main road and up a cutting and the big tussocks, right overhead, and John and I were in the dickey-seat in the back – ducking down to avoid the tussocks. The road was cut down and these tussocks were right over the top and we just disappeared into them, like into a tunnel.'

At least there was already a house and station buildings at Ngamatea. John Spear, who first knew the place in 1915, described them as being 'out in a vast unrelieved expanse of flat and rolling country, with no sheltering timber except a young plantation surrounding the homestead, a collection of whares pulled together, lacking in many modern conveniences, but at least very comfortable. The woolshed, cookhouse and men's whares were some distance away by the side of a large creek.'[9] Drummond Fernie's timber was soon put to good use. Gordon Mattson and Lou Campbell, both now in their nineties, were there in the mid-1930s. 'Chips', said Lou, 'stood out from everyone else. His personality was totally different. He

Johnny and Noel Roberts, 1931. 'The cookhouse, now lined and spruced up minimally' became their first home at Timahanga.

kept to himself – whether he had a wife or what I don't know. I never heard of him talk of any of his connections at all. He was a first class tradesman. Everything there seemed to be connected to him.'[10] Gordon remembered Chips 'renovating' the existing ten-stand woolshed and 'always adding bits to it. He made a marvellous job of it.'[11] He certainly extended the woolroom and when the shed was finally replaced in 1984 its history was revealed: the beams under the floor, with the names of long-gone shearers and shed hands carved into them, apparently came from the old Studholme woolshed. It had always been believed that when the 20-stand blade shed was demolished some of it was used to build the shed at Ngamatea. The stables, especially, were a work of art, praised by generations of musterers for their design and construction. They were huge – ten loose-boxes inside and separate stalls for the packhorses outside, doors at both ends, and a big chaff loft upstairs. The blacksmith's shop and implement shed were another of Chips's creations, and in 1938 he built a new cookhouse and shearers' quarters.

The Ngamatea homestead had begun life as three whare out at the Tikitiki, part of Renata Kawepo's 'headquarters' or the 'outstation' marked on early maps, where there was a woolwash in the early days. Until quite recently, the outline of the foundations in the tussock were clearly visible from the air. In the late 1870s or early 1880s, the whare were dragged in from the Tikitiki by bullock team, and 'put together' at the Ngamatea

Ngamatea homestead – an idiosyncratic house which began life as two or three whare at the Tikitiki bush.

station headquarters. It was the first homestead in the area. Over time a room or two was added here and there, but there was a hole in the middle of the house where two sections met. In winter it would fill with snow and freeze. 'Noel and I were playing around on the roof one day,' John Roberts recalled, 'and he slid down into the gap between the bedroom on this side and the dining room on the other. It was a big space – a couple of feet. I had to get a rope and pull him out.' This idiosyncratic house was clad with corrugated iron, 'very few windows were the same size, the doors were different sizes and some of them were made of only about three or four slabs of timber adzed out of the log. Some of them were probably 50 millimetres thick – three or four adzed planks with two or three cross members and that was your door. The door that went into the front room – the sitting room – that was adzed.'[12] Despite its unlikely beginnings it became a warm and welcoming house, a nurturing place, the heart of Ngamatea.

When Jack and Dorothy Roberts moved in they soon took down the pines that were growing right up against the house and let in a bit of sun; John Spear's 'young plantation' had matured. Both parents were fully occupied, Dorothy with station work as well as running the household, so the boys were given their freedom and enjoyed it while they could. They made the most of their new playground with their ponies and trolleys and watching the seasonal round of farming activities. They spent a lot of time with Tom Whimp, the great station blacksmith.

Jack Roberts and Johnny at Ngamatea.

He was a crackerjack. We'd be making carts out of disc-carrier wheels so we'd go down to Tom and ask him to weld this up – all blacksmith's welds. We should have killed ourselves with the things. There were old stables at the homestead, just along a bit from the store, and we used to drag this cart up the hill and get in. It had an old tea-chest on it and we'd sit there with our heads poking out, and come down this hill and then we had to turn sharply to get into the stables. I remember one day coming down and there was a horse in there. We didn't know and we hit the gates of the stall.[13]

When the shearing gang arrived there were other kids to play with. Once they were playing hide-and-seek and Noel thought he would hide down between the bales stacked high in the woolshed waiting for a truck that would come up with fertiliser and cart the wool back down to Napier. 'I slid right down to the bottom onto the floor, from three high. Well, how was I going to get out of here? The kids went and told the pressers what had happened, so they got a wool-hook and tied two or three lengths of seaming twine onto it and lowered that down. I remember grabbing hold of the wool-hook and they hauled me up.' The end-of-shearing dance was held in the cookhouse, and John remembered the men putting a bit of chaff on the floor and pulling him and Noel around on a sack. 'That floor was really fast, and it was amazing the musical talent that came out. I can remember one dance, there was an old swagger who happened to be there at the time. He got a dozen beer bottles, filled them up to different levels with water, put a teaspoon in each one, and he had a xylophone. There was always a saxophone and guitars and ukuleles, and they brought the piano down from the homestead on the truck.'

For eighteen months in 1932–3 they had a Miss Jacobi as a governess or nanny. She was very prim and proper and had never been on a farm before.

> She would take us for a ride every day and she'd sit up there, real jockey fashion. Noel would say he'd go ahead and open the gate, and he'd duck behind the scrub and we'd just ride past and she'd think he'd gone on. Then I'd open the next gate and I'd duck off, and just leave her to it. Noel and I would come home, and oh, the consternation when we got back. We'd get hauled over the coals. Walter Fernie was there one day, and he stood back and opened the gate for her – she was sitting up so prim and proper – and he said, 'Many a good horse she hasn't ridden.'[14]

Noel was ready for school but his mother had no time to supervise correspondence lessons and Miss Jacobi does not seem to have provided any teaching. He happily avoided formal education for two years until he turned seven and was shipped off to the Friends School in Wanganui as a boarder.

Timahanga and Pohokura were now outstations of Ngamatea. The whole property was about 70 miles long and 30 miles wide. There was already a whare – a cookhouse – at Pohokura, built in the late 1800s of pit-sawn timber from the nearby bush. A cook and one shepherd were stationed there, and the precious merino rams the Fernies imported from Australia were kept down on that lower warmer country. The rest of the shepherds would go down for mustering and docking, and at the end of the day they could relax in Pohokura's hot spring. 'It was a dug out hole, in the clear, not the bush, but if the wind was blowing in the right direction you could smell the sulphur. It was only like a puddle and you could get into it over your waist. We used to lay back in it, but if you stopped there too long the fumes got you, and you'd go to sleep.'[15]

There were still big scrub-cutting gangs at work and Drummond Fernie set out to buy the equipment he needed for the next stage of his development programme. He looked an odd character in town in his old greatcoat

The uncashed cheque found in the Fernie Brothers papers in the 1970s. 'Every time a cheque came in Drummond would stuff it in his pocket and forget about it.'

and working pants and boots, and with a sugar sack full of money tied with string over his shoulder: he preferred cash. It is said that one of the men found a coat of Drummond's draped over a fence, going green with mould, and when he went through the pockets each was full of uncashed cheques. 'Every time a cheque came in Drummond would stuff it in his pocket and forget about it.'[16] It sounds an unlikely story, but years later, when all the Fernies had gone and their papers were being sorted at Otamoa, an uncashed cheque on Dalgety & Co., dated 8 April 1926, came to light. It had never been cashed. It was for £795 12s, the proceeds from the sale of 600 ewes.

Drummond still had enough money to go to Palmerston North to shop for tractors and he wandered into the showroom of Gough, Gough & Hamer, the agents for Caterpillar. He was known there, but not by the young salesman who attended to him and soon got fed up with this unlikely-looking customer and his endless questions. Suddenly Drummond said, with his characteristic stammer, 'I say, I say, I'll have tttwo ttractors, the bbig one and the small one, and two pploughs, and ttwo setts of ththose ddiscs and harrows.' The poor salesman excused himself and went for help. When the manager recognised this old codger who wanted to buy the place out, he scolded the young assistant and made a great fuss of Drummond – 'Hello, Mr Fernie, hello, how are you' – and hastened to write out the shopping list. The tractors bought that day were a Caterpillar 30 and a Caterpillar 20. Noel Roberts remembered that 'Gough's had the 30 in their showroom for years – took it back from Ngamatea when they got a bigger one – sort of trade in. She was really antique. They were the first tractors Fernie Brothers had at Ngamatea.'

Of course Drummond needed swede seed and fertiliser too and he got his crops in the ground, hoping they would do well up there. Everyone knew the Fernies, or knew someone who did, and they all took an interest in what was going on. When an acquaintance saw Drummond in the street one day and asked 'How are your swedes doing Drummond?' he replied, 'They're ccooeeing tto each other.'

The scrub-cutting went on for years, not just at Timahanga and Pohokura, but on Ngamatea as well, where they cut part of White's and then the Gully block. The station hands were kept busy packing out the tents and other gear. The cut scrub was left to dry out for up to a year, then burned. Noel remembered a big scrub fire at Timahanga when he was eight or nine and home from school for the holidays.

> Dad got all the scrub cut at Timahanga, Jimmy's and Pohokura – ten gangs of Hindus from India – dinkum Hindus with turbans, about a dozen men in each gang. They'd come into the country for the purpose. I was in the cookhouse and the fire was just over the creek. I put my hand on the glass and burned it. It was an intended fire – lit in several places. There was cut scrub just over the creek so it came that close. Seth Barnes was overseer at Timahanga at the time. I remember him packing stores, grass-seed, posts out after the burn.

They began cultivation on Ngamatea too, sowing 'the Long Oat up towards Oat Paddock hill'. Several hundred acres of oats had to be grown each year for chaff for the packhorses used on the musters of the back country. It was September 1932 when Tom Whimp joined the team. He was both blacksmith and wheelwright, 'a very versatile man. He could make anything – all by hand, in the forge.' His greatest piece of handiwork was a triple-drill hitch to tow three drills at once – 'all seventeen-coulter drills – that's the widest they made in those days, and you pulled the three of them behind the crawler tractor'. With this clever invention they could drill and sow down 100 acres a day. 'Tom made all the gates too, of hardwood, with iron hinges and gudgeons and fasteners. He made so many things, although shoeing was the main job.' It was a real bonus having a wheelwright on the property. He and Chips, the carpenter, made a dray and four-wheel and two-wheel wagons needed for carting oats and hauling firewood in from the Tikitiki bush.[17] The little wagon is still at Ngamatea. It came to grief in the Otupae block and sat out there for ages until it was rescued and brought back to the homestead, where it is now a showpiece in front of the cookhouse.

Tom Whimp's handiwork, a triple-drill hitch. They could drill and sow down 100 acres a day with this clever invention.

In the early 1930s several of Ngamatea's great characters arrived on the station to make war on rabbits. The first, Peter Duggan, may have been inherited from the Ruddenklaus; he appeared in the wages book on 10 March 1931 and stayed five or six years. Initially, in the depths of the Depression, he was getting 30s a week. He had a daughter in Wellington but as he could neither read nor write he used to get the gardener to write to her, then read out her replies. Gordon Mattson remembered him as 'a happy old chap'. His name is memorialised in the landscape: Peter's hut, Peter's Ridge, Peter's Top – usually just Peter's, which others may call Tauwheketewhango, or Tawake Tohunga.

Jerry Johnstone,[18] who started a week after Peter, was around for years and remained a friend of the Roberts family. When they went to his eightieth birthday party at Otairi station, near Hunterville, 'he still rode thoroughbreds and still gambled. He was still holding down a job, this time as storeman, and he handed out tools, gear and sarcasm to all the younger generation.' They kept in touch with him when he went to the Home of Compassion in Wanganui where the nuns cared for him and 'kindly listened to his advice on racing, and what horses to bet on'. Winnie Roberts, in her later years, wrote down some of the stories about their old friend:

> I'm always sorry I didn't take a snapshot of my friend Jerry. Perhaps it
> wouldn't have revealed much of his character. There were many do's and

don'ts attached to one's dealing with Jerry. Impatient, sarcastic – he didn't suffer fools at all, yet generous to a fault. He would work months at a time, then go off to town and come back broke. I remember one time he had a big fat cheque and he was going back to Australia; he said his good byes to the ones who mattered to him and departed on the mail truck. Poor Jerry, he was back in a few weeks – broke, ready to start all over again. He had listened to many tales of hardship – not all genuine, I thought. 'Poor little thing, what else could I do but give her a handout; he deserted her and the kid, the dirty so and so'. He was fair game for children: ice creams for the whole gang who collected round him.

And Winnie, too, benefited from his generosity. 'On Sundays he would bring his thoroughbred mare bridled and saddled up to the house: "Thought you might like a ride on a real horse for a change". I had the choice of any or most of the station hacks, my own saddle and bridle, but it gave him great pleasure to have me ride off on his beloved mare.' As one of the rabbiters, Jerry shifted from camp to camp; he spent long periods at the Golden Hills hut. Gordon Mattson, who had arrived at Ngamatea as a very young cowboy,

> wanted a 'go' at living in the Kaimanawas with Jerry, and on weekends when he came in for stores and mail, he regaled us with stories. Jerry was a good cook and tried his hand at bread, pies, 'duffs', etc; a real good camp-oven cook. One evening when Jerry endeavoured to hook out his masterpiece of plum duff, the cloth slipped off the fork and the duff fell back into the billy of boiling water which came up like a geyser and on the downward turn over the top of poor Jerry's bald pate. Gordon was aghast. Miles from anywhere and Jerry badly scalded. Jerry danced around. 'Quick, quick, get the oil' he yelled. These old hands seemed always prepared for emergencies. Lime water and olive oil mixed is still a great healer of burns, and Jerry had his bottle planted in a corner of the whare. A liberal application poured on, his head then swathed in a towel, and Jerry looking like a Sultan took himself to his bunk. In the morning when Gordon awoke he was in a dilemma. It was an unwritten law to anyone living with Jerry not to speak first in the morning. Wait for Jerry to open any conversation. So Gordon waited – and waited. It seemed churlish not to enquire, after what had happened. 'Perhaps the old chap will feel neglected if I don't ask. Well, here goes' – and Gordon brightly asked 'How's the head today Jerry?', and Jerry bounced to a sitting position, eyes popping, his turban awry, and snarled 'How the hell would I know'.[19]

George Everett, liked by all, lasted longer than any of the other old-timers. He started at Timahanga in November 1933 and was there until he disappeared from the wages register in December 1936. He reappeared in June 1943 and stayed until he retired in August 1968; he made the station his home. He was well educated, well informed and came from a good family; beyond that there was a great deal of speculation, although most agreed that way back, when he fell out with a girl, he just decided to go bush – and stayed there. He did not talk about his past and there were so many people at Ngamatea with doubtful histories – and often the most unlikely names – that one did not enquire too closely. But George was a most remarkable man: tall, strong, a hard worker and a great walker. He could approach a fence carrying a huge log on his shoulder and without stopping or dropping his load just stride over the fence and keep going. If he had to take a horse anywhere he dragged it along behind him. If he needed something from town he walked in to Taihape and did his shopping, and walked back to Ngamatea – 35 or so miles each way. One day as he was coming up the hill from the Rangitikei he heard an old Model T come down the Springvale hill towards the river then start chck-chck-chcking up the hill behind him. He was nearly at the top when it caught up to him and stopped to ask if he would like a lift. 'No thank you,' he said, 'I'm in a hurry.' It was not far from where he would leave the road and take a short cut across country and down to his camp at the bottom of White's block, but he did not want the company and he did not need the ride.

George was rabbiting on Te Awaiti in the southern Wairarapa before he came to Ngamatea, and it was said 'he kept his camp absolutely immaculate and had a big team of dogs that would eat anyone who came into the vicinity if they weren't careful'. George expected Te Awaiti would be cut up for closer settlement so he left to seek more open space, and found it at Ngamatea.[20]

Gordon Mattson started at Ngamatea in October 1934 as a general hand, later foreman, and was there until the end of 1946, so he knew all the old-timers. A keen rider, he knew the place well and loved every part of it. He had worked for five years on a property near Otane and left looking for a more congenial job. He found it through the Labour Bureau in Napier. 'I worked with a gang of men up there. Tau Wilson was foreman and there were several other Maori men, great workmates.'[21] Winnie Roberts thought well of Tau Wilson: 'he was a good fencer, fair carpenter, a big strong man of character and stature. He decided when the hay was to be cut, when the oats were ready, and he was the stacker who built the lovely symmetrical stacks thatched with rushes. Tau married a very fine widowed lady who ran a shearing gang and he came back to Ngamatea many times as boss of the gang.'[22]

The only vehicle on the place was the boss's Chrysler 70. John Roberts remembered it: 'a coupé with a dickey-seat, that used to crawl round all over the place with chains on. Mum would put the chains on, and take us kids the 7 miles out to the road. It used to take a couple of hours.' And when they were cutting oats with the reaper and binder and the tractor broke down 'they hitched the old Chrysler on. They had to stop now and then because it boiled, but they got the job done.' If the Chrysler did not go out for the mail someone would ride out for it and this was often Gordon Mattson's job. He would avoid taking a packhorse if he could and hope he could make do with a pikau, a makeshift saddle-bag, but inevitably he would get caught out and he would have to load up the horse and walk back in leading it. As it still is today, the mail-box was 2 miles down the Taihape road, at the Otupae turn-off, and it was 7 miles from the Ngamatea gate in to the station. Sometimes there would be an unexpected passenger on the mail van too. 'One time there was a woman with a little child and her husband who was a scrub-cutter. The woman was on my hack with the baby, and her husband and I carried in miscellaneous goods and what have you.' Sometimes, if the shepherds were in at the homestead, one of them would go out for the mail. One day it was Lou Campbell's turn.

> This chap arrived on the mail truck. Apparently he didn't let Jack Roberts know he was coming. He never had a suitcase – he had his belongings in a canvas bag. I didn't have a horse for him and I took his swag and he walked in. I said something like he'd be lucky to get a feed or get a job, and he said 'Oh I think I will work that one out all right', and it wasn't until next morning, he came down to meet the boys, and I got the shock of my life. He was Walter Fernie. Apparently he just used to turn up. I remember Jack Roberts more or less saying he was a bit of a nuisance because everything had to work around what he was doing.

Drummond Fernie rarely visited Ngamatea, but Walter used to drop in now and then; he enjoyed nothing more than getting out around the place, and especially rounding up a few wild sheep. He relied on others for transport: he had had an accident in the early days and never drove again. 'We don't know what happened, we never got to the bottom of it, it's just hearsay, he wouldn't talk about it. Helen, his wife, used to drive him everywhere, through the night, in the early hours of the morning, dogs in the back in the three-seater De Soto. She would drive him to Ngamatea – she was a great one. Eleanor, his second wife, didn't drive, so he had to get some of the men on the farm to drive him around.'[23]

In 1935 or early in 1936 the station bought a brand-new 3-ton Chevrolet truck – a real luxury, though chains were still needed more often than not. The station road was like a great trench, scoured out by wear and weather and filled with snow in winter, and in the early days there were fourteen or fifteen gates. If there was a truck or tractor going somewhere Johnny and Noel were always around. 'I think the men were glad. Mail day one of us was always there, opening and shutting gates.' But there was an alternative form of transport. Jack Roberts had a friend called Gerry Durand who was chief instructor at Hawke's Bay–East Coast Aero Club, and when Jack had business down in Hawke's Bay he would call Gerry to come up and fly him down in his Gypsy Moth. Once, down in Napier, Gordon Mattson missed his ride back to the station, so he called Gerry to see if he could fly him back to Ngamatea. It was going to cost him more than a week's wages – but his step-brother gave him a hand by doubling him out to the airport on his bike. 'We landed at Ngamatea. Did a circle over the house first and got lined up for the aeroplane paddock and Jack Roberts came walking down to meet the plane, thinking this was some important gentleman coming on business. And it was me.'

In 1934 there was a medical evacuation by plane from Ngamatea. Dr David Bathgate, a keen tramper who knew the Roberts family and Ngamatea, got a call one clear frosty winter afternoon to attend to Dorothy Roberts who was expecting twins and needed urgent medical attention. The station road was half blocked with snow and the patient too ill to face a tractor and trailer ride, so Gerry and his Gypsy Moth were called in. All he needed was a good smoky fire at the top end of the 'best' paddock for him to land in. The station accountant, 'a charming young lady', arranged for the smoke signal and the plane made it in through the snow and mud. The doctor decided against immediate evacuation of the patient and settled her down for the night. She would have a good sleep and the hard frost would facilitate take-off early next morning. 'We lifted the patient into the front seat for her first plane ride. Gerry revved up the engine, then the wheel chocks were removed and Gerry took off down the frozen paddock with full throttle. Down towards the lower end Gerry lifted the Moth over the fence and the adjacent gully, climbed up into the sky and did one farewell circuit round the paddock, then headed the little Moth up into the long climb over the ranges.'[24] There was no room for the doctor on the plane: he had to help dig a track through the snow-drifts so they could haul the car out and drive him back to Hastings. The patient survived the journey and a couple of weeks later 'presented her proud husband with twin boys', red-haired David and fair-haired Barry. Noel and Johnny were there on the aeroplane paddock when their mother and the twins arrived home by plane.

Ngamatea's 'charming young lady accountant' was Winnie Craven, who had gone up to Ngamatea at the beginning of 1934 to give them a hand for a few months. She was there for 33 years, 'the happiest times of her life'. Her job did include doing the station books, but she also ordered the stores for the whole station, including the Indian scrub-cutting gang,

Winnie Craven supervising Johnny Roberts's Correspondence School lessons on the verandah at Ngamatea homestead.

and then delivered them down to Timahanga once a week. She ran the store at Ngamatea, was governess to Johnny Roberts and supervised his correspondence lessons, and lent a hand with whatever else needed doing. She loved Ngamatea from the start – the miles and miles of unfenced tussock, the vastness and the grandeur. 'I remember the first time I saw this country – the sun was shining and it was gold for miles, it just looked great, and I said then I believe I could put my roots down here.' It was great horse country and horses were a passion: she bred them, she trained them for the Hawke's Bay Show and won prizes with her harness horses, and she drove for Ted English, a well-known huntsman and horseman from Meeanee, in the ladies' events at the show. Winnie could ride all day; she was in her element at Ngamatea. One day Jack, Dorothy and Winnie left the homestead to ride around the station boundary – as near as they could, given mountain faces and river gorges. They only got so far and Jack's horse put his foot in a rabbit hole, 'and that was it'. It would have been a long ride.

Gordon Mattson shared Winnie's enthusiasm for horses, and they often rode together, once over the Otupae Range to Pohokura. It was pretty high, and the first time Gordon had been over.

> We went up from the northern end of Otupae Range, fortunately, and along the top to the telephone line and straight down the edge of the bush

to Pohokura. We got about three-quarters of the way down to a gate and I happened to see a hind get up and off for its life – and there was a freshly born fawn. We put the reins over the horses' heads and gave them a slap on the back to go down the fence to where the gate was. Winnie wanted me to take a photo of her with this fawn. I took it and we were going to continue on, but out of the corner of my eye I saw my mare had turned round and started to go up again, and Winnie's horse was following. I hadn't a hope of catching them up – it was as steep as anything and I had leggings on. No time for dilly-dallying – I left Winnie sitting there, and I off, straight up. Fortunately we'd taken the end route over the range, and I thought, Those horses will take the same way back. So I went straight down the old track, which was the middle of the range – straight down the telephone line. I just got to the bottom puffing and blowing, and there were these horses, just coming back down. If I hadn't got them there, they'd have gone all the way back to the station. I hopped on my horse, and led Winnie's, and I made them go straight up again to the top. Winnie meantime had got very worried, so she in her good time had got to the top and was waiting there. I've never seen a person more pleased and relieved to see me and the horses again.

Lou Campbell remembered Winnie riding out to Peter's hut to take voting papers to the musterers, but for what election he did not know. Another time she gave them a hand to muster young horses in from the Swamp block at Ngamatea. They had been turned out when they were weaned and left there until they were almost mature.

It was a huge area, with some very bad patches of flat, spongy swamp in the middle, and it ran the whole width of Ngamatea between the Rangitikei and Taruarau. They decided to round them up and get them in, and we had about three goes at it. Winnie came out with Jack Roberts and us about the second trip. First go they just beat us because they knew their way through the swamp, and in the finish Jack had the idea to hit them from both ends, because they used to just run from one end of the block to the other, and beat us. I actually caught one of them – a mare heavy in foal – I put a rope over her. In the finish we took out one or two old station horses and let them run with them and that didn't work – the station horses came home and the others still went round. It was a real ordeal, just about everybody got knocked up at it, and I don't know if we got them all or not.[25]

Winnie was as handy around the homestead as she was out on the station. It was an important job ordering the stores, coping with the mail and doing the books, and then there was Johnny to cope with.

I was a bit of a lad. Winnie was in the bath and you had to go through the bathroom to get to the toilet. I knocked on the door and called, 'Quick, hurry up, I want to go to the toilet'. 'You can't – I'm in the bath.' 'But I want to go, I want to go, let me in.' 'Oh, all right. I'll unlock the door – but you let me get back into the bath and when you come through you just go straight to the toilet.' I waited. 'All right, you can come in.' I came in, and instead of going straight to the toilet I went up to the bath, put my hands on the side, and looked her up and down and said, 'Gee Win, aren't you a little corker.' At that age! We got such an education into the world and the ways of men, and we were just ordinary kids, we just accepted everything. That was life.[26]

With Noel away at school Johnny was short of company until he too turned seven and went to Friends School. One day he was drawing Jerry Johnstone's dog Booze for Correspondence School. He drew a big dog and painted it bright orange with black stripes. 'But it looks like a tiger' said Winnie. 'That's all right. There's tiger in him. There's lion, tiger, monkey, elephant and pelican in Booze.' 'How do you know that Johnny?' 'A man said, "That's a fine dog Mr Roberts – what's his breeding?" and my dad said, "A bit of everything."'

Booze and his sister Peggy were well known on Ngamatea. They were lurchers – 'greyhound crossed with a heavier dog, with a smooth coat, short hair and strong bones', according to Gordon Mattson. Lou Campbell remembered the dogs too. It was Jack Roberts's policy that nobody was to be camped out in the bush on his own in case of accidents so when Jerry Johnstone's mate was to be away for a few days Jack sent Lou out to stay with him. 'Then it was Jerry's turn to go away and I looked after Peggy and Booze for a day or two and you couldn't put your hand on them. They'd growl at you and just about snap your hand off – I just threw them a bit of tucker.' There are, though, photos of Noel and Johnny playing with Booze and even riding on his back.

It seemed to Winnie that 'it was a cold cycle' in her early years up there. The house was at 3,000 feet and they would often be snowed in. 'The snow showers started in April and continued . . . well – I've known snow on Christmas Day, and I've known snow on New Year's Day. It was a very varied climate – we could get some very hot weather. Many winters the mail van couldn't get near us, but when we had the crawler tractor it would go 20 miles to meet it to bring our mail in.'[27] Before they even got the crawler tractor out of the shed they had to break up the frozen mud under each track to get it started, and then they had to be careful as snow would build up underneath it and it could slide. But you got around one way or another; there is a photo of the twins in the snow, with dog chains wrapped round their tricycle wheels.

Johnny and Noel Roberts with Jerry Johnstone's rabbit dog Booze, a lurcher.

Over the years Winnie quite often had to deal with the police. Her son, Jack, remembered how they would ring up and ask whether someone was on the pay-roll, and she would check the wages book.

> If he was, the cop would say, 'When he leaves, or comes to town, can you just let us know.' In those days some of them were dodging maintenance, and you paid your income tax once a year and everyone tried their best to dodge that. On this particular occasion Sergeant Chestnut, from Taihape rang up and gave Mum a list of six names and Mum said, 'Your luck isn't holding very well this week – we've only got five out of the six.' If they weren't doing anyone any harm the cops would leave them there till they decided to move on, then they'd pick them up when they came into town.

For a couple of years from about 1935 Ngamatea had its own football team. They played Mangaohane and maybe other stations, and the Huia club of Taihape, but they only played 'now and then' when a team had a bye and was free to play Ngamatea. They had hardly any practice, but their jobs kept them fit. Their captain was Bob Baylis, who had taken Seth Barnes's place as head man at Timahanga, and Gordon Mattson, Lou Campbell and George Everett all played in the team. Gordon was 'a sort of roving wing – all over the place'. He named some of his team mates: Ted Isaacson, George Stevens, Bernie Doole, Noel Burnell and Alf Murphy, 'among others'. Sometimes Gordon would take Johnny and Noel with him to see the footy. 'We used to run up and down the sideline with a bottle of smelling salts. If anyone got knocked out we'd wave this under their nose. One day George Stevens was late – he was a great man for the women I suppose. Noel and I were on the sideline, and he arrives at the field with a girl on each arm. He gave his top teeth to one, and his bottom teeth to the other.'

Of course the after-match function figured large. Sometimes there was a dance, maybe down the road at Winiata, but as often as not the Gretna was the scene of their revelries and Jack Quirke, the proprietor, had rooms at the back of the hotel 'he kept for footballers'. Taihape businesses looked after the boys: the stations and their staff were good customers. On one occasion they had been in to Taihape to play Moawhango and it snowed on the way home. They were all on the back of the truck, apart from two or three smaller ones in the cab, 'and there was a tarpaulin we had to get under but George Everett, who was over 6 feet tall, was taking the weight of the tarpaulin on his head and we had to stop and get into a paddock at Erewhon and get some branches to hold the tarp up'.[28]

In the Depression years various travelling salesmen used to visit the stations on the Napier–Taihape road. One of them was Jack Lager, whom Gordon Mattson remembered well:

> He was an elderly man, an ex-steeplechase rider. He was an uncle to a men's clothing firm in Napier and he had an old two-horse buggy he used to drive right up from Napier and visit all the farms and stations round about selling men's clothing to the station hands. Jack was quite a character. Apart from having good quality clothing he was skilled at sharpening razors – the station hands would always bring him their cut-throat razors to sharpen. He was also quite an entertainer, a ventriloquist, and he would put on a bit of a show for us, which was very good. He used to come about every six months with his goods and stay the night and as we bought stuff off him he wrote our names down and the article and the cost and when he was ready to go the next day he'd take this up to Jack Roberts. Jack would write him out a cheque for the total of all our purchases and away he'd go – and it would all be recorded against our names in the wages book.

Stock numbers are not available before the 1937–8 shearing when just on 16,000 'Ngamatia' sheep were shorn – and it is not clear whether that total included Timahanga. Noel thought 'Ngamatea was running about 12,000 sheep when Dad was there. He was there eight or nine years, and he could have built the numbers up.' Lou Campbell believed the tally was higher: 'It was a good number considering there were only 5,000 acres, the freehold block, in grass or under cultivation'. After all the scrub was cut between Timahanga and Pohokura in the early 1930s and the land was in grass, they got the ewe tally on those blocks up to 10,000 but the numbers kept dropping after that as the scrub came back. In those days they lacked the means – fencing, fertiliser, machinery – to keep the scrub under control. On Ngamatea, the old Owhaoko, they grazed many of the blocks that the Studholmes had grazed, but sheep numbers went back as hard grazing brought about the gradual replacement of the original soft tussock by tall, less palatable snowgrass, which in turn gave way to a short bristly tussock. The Fernies brought sheep up from their properties on the Parapara: there was no market for them during the Depression and it was 'a convenient place to park them'. There were no boundaries except the rivers and they could push the sheep back as far as they liked even though there were no formal leases over most of the back country. The Fernies were graziers, fine-wool men; like the lessees before them they ran halfbreds. There were still numbers of wild sheep around – descendants of Studholme's merinos and others which came in by Mount Cameron or the Ruahine Range – and there had not been a clean muster for some years: there were hundreds of stragglers out the back – sheep that had been missed in the annual shearing muster.

Lou Campbell went to Ngamatea in February 1936 as a young married man. He had already put in some months at Mangaohane, and there was no married accommodation on either place so his wife Joan stayed with her family down in Hawke's Bay. A job was a job, and Ngamatea was a huge challenge. 'All he could talk about was Ngamatea and the size of it and all that. It was a big place for him – he was only twenty-two.' Lou did part of two seasons up there plus the winter, a particularly hard one, in between. Phil Brown was head musterer and there were another two musterers – Stanley Doy, 'a little Englishman', and Ted Isaacson. They were not around the station much but 'were out at the huts pretty well all of the time'.[29] This was the pattern of the muster at Ngamatea, as it had been since the earliest times, and it would continue through the 1960s, as long as they grazed the back country.

Jack Roberts would go out on the muster with them. He was 'rather short-tempered, and a sick man – you could smell the insulin he injected into his hip somewhere – and of course he stuttered'. There was no packman-cook, and some of the wily old packhorses had come with the station and went back many years. The muster was hard work with so many hermit sheep that had missed several shearings, and were dragging around several years' growth of wool. Lou and the others were coming from the

Kaimanawa hut one time and mustering off a peak known to the musterers as 'the Michael' but more correctly called Whakamarumaru.

> Actually we got the sheep high up. Wild sheep or those left on their own, graze down and then they work back up – especially anything with merino blood, they work uphill to camp. We'd have got perhaps 20 sheep. There wouldn't have been any more. To get them out we had to trace clip them – part-way up their sides. There were two or three streams to cross on the way, and once they got wet, you couldn't do anything with them. We got bits and pieces all over the place that had escaped and learned to hide back. Once anybody's had a go at mustering the sheep in the wild they get cunning, like deer or anything, and they can pick up any noise.[30]

There were cattle as far back as the Boyd that had been out there for a year or two before Lou's time. They picked up some of them, lost them in the bush in a snowstorm, picked them up again back where they had started, and eventually got them into a holding paddock at Golden Hills, then back to the station.

Another time, down at the southern end of the station, they were mustering out Blowfly Gully which drains the top of the Otupae Range and 'gets very gorgy as it goes down into the Taruarau, very cliff-like on either side. It was a hermity sheep hold-up in there. If they heard a dog bark they'd be over the bank and down onto the roughest part.' There were three of them on the job – Phil Brown, Lou and Ted Isaacson – and things did not go smoothly. They had got the sheep across the Taruarau but could not get them any further, burdened as they were with their massive wet fleeces. It was getting late and Lou found himself alone on the job – the other two must have called it a day and gone back to Timahanga. He was about to go too when Jack and Lawrence Roberts rode over the skyline, expecting to give a hand with the muster off. 'Lawrence used to come up there and visit now and again – he was probably courting Winnie at the same time.' Neither Jack nor Lawrence could see any point in going back to Timahanga for the night: they just prepared to camp where they were. 'I never forgot it,' said Lou.

> It was one of the coldest nights. I camped with them – on the top, in the open. No blankets or anything. But Jack Roberts, once he'd given himself the injection that he used to take with him all the time, he could doze off and go to sleep. He was snoring his head off and I couldn't get to sleep for him snoring. Next day they came out with food for us from Timahanga – we had nothing with us – and we carried on and got the sheep off. There was a chap, Barney Scott, who was down at Pohokura and he came out and joined us, and he was noted for his good heading dogs. He helped us bring them in.

Winnie once said that the first time Lawrence asked her to marry him she said, 'Oh, Lawrence, ask me again in ten years.' And he did.

> The day came when the news went round the station I would be leaving. I was to marry the boss's brother and live in Hawke's Bay on the station he managed for his uncle Drummond. Jerry was one of the first with his congratulations. He came up to the house before breakfast, looked in the living room window on passing, and saw me setting the fire – at that altitude a daily necessity. He opened the front door, tiptoed in and knelt down beside me, put his arm around my shoulders and said, 'Congratulations, I've just heard the news and I'm so happy for you. You'll never have to work again!'. And all my married life when I was extra busy, the days too short for all the work that had to be fitted in, I would remark to an equally busy husband 'I wish I could tell Jerry about this, he was so pleased I'd never have to work again'.
>
> The boys gave me a party in the cookhouse before I left. Jerry had asked me what would I like for a wedding present and I was afraid his generosity would run away with him, so I asked for a deerskin mat for my bedside. Everyone on the station shot deer and cured skins, so I was keen to have one in my new home. I asked the boss to refuse writing out a cheque for Jerry until after my departure. He circumvented that one, and when at the party the boys presented me with a lovely china cabinet – even the Hindus had sent in their donations – Jerry came in with his gift: an oak writing desk, both pieces being smuggled in to the station with great secrecy. It was a great party. Everyone contributed his item with enthusiasm, and cheers from the others.[31]

Winnie and Lawrence were married on 28 April 1937 at the Methodist church in Hastings. They settled at Mangatapiri and expected to be there for life, but fate intervened. Jack Roberts decided to take a cruise to the islands, a health cure, but the weather was rough and he got terribly seasick, could not eat and because no one on the ship knew he was diabetic, failed to get his crucial injections. On 19 June 1937, aged 38, he died and was buried at sea. Dorothy and her sons stayed at Ngamatea for some time before they moved down to Hawke's Bay, and the boys always went back to their old playground for holidays.

In true Fernie fashion Drummond changed around the ownership of all the properties after Jack's death. He made Lawrence his partner in Ngamatea and Timahanga and signed Pohokura over to him. At least he left them a few more months at Mangatapiri – until after the birth of their first child, John Ramoth Roberts. They moved up to Ngamatea in April 1938. Jack junior was just six weeks old. Then in December 1940 Margaret Catherine Roberts was born, and the family was complete.

Chapter Three

Lawrence and Winnie

*These people were the salt of the earth,
the backbone of the country.*

When Drummond Fernie died in 1940, his younger brother Walter became an owner in Ngamatea and Timahanga and partner to Lawrence, and this combination was to last for 26 years. The Second World War had begun in September 1939 so most of the younger men – the shepherds – and a few of the old hands were soon called up or manpowered. Tom Whimp was manpowered down to the Addington railway workshops in Christchurch, but Colin Kirkpatrick, the marvellous gardener who had started in August 1937, stayed on – for 20 years. Colin had lost his wife in the Hawke's Bay earthquake 'and it ruined his life. He came up to the backblocks for solace. And he found it.' He had 'an incredible garden, a couple of acres, and grew all the vegies for the house and the station – everything, even strawberries which were under a big netted frame. Every Friday he'd send fresh vegies down to the cookhouse at Timahanga and often there was a surplus which would be bagged up and sent in by the mail truck to Taihape hospital.'[1]

Then there was Englishman Bert Jeffery, known to all as Jeff. He started out the back of Ngamatea as a rabbiter working for the Rabbit Board and in 1937 he came to the station as a general hand, remaining there for twelve years. Jeff was a bit of a hermit, but he was not a remittance man, hiding out on the other side of the world, and he kept in touch with his family. A plumber by trade, he used to do all the odd jobs around the place 'if you could talk him into it, but he was so bloomin' contrary. Not sure if he was a drinker – think he might have been – most of them were. He was an intelligent man – wrote a lot of poetry – but he'd gone astray.'[2] Jeff was born in the same house as Sir Francis Drake and, as Winnie said, 'he didn't let anybody forget it either . . . . And he ended up at Ngamatea like a lot of other deadbeats, but he really was a character. He was clever in many ways. When I wanted things done I would appeal to him – like new drains. "Yes," he said, "I've got a drain-layer's certificate", and they went in beautifully.'[3]

Gordon Mattson spent some months out at Golden Hills with Jeff, poisoning rabbits and shooting wild sheep across the Ngaruroro. That was their mutton supply.

> He was well-educated, old Jeff, but a proper grouch – and his weakness was grog of course. He'd stay at the station for twelve months and then he'd go into Taihape and have a real burst. But the pity of it was his cronies would always wait for him to come in and give him a hand to relieve him of his cheque. Both him and Jerry Johnstone were the same – they used to go straight out to Hihitahi and get on the train and bypass Taihape to dodge these hanger-ons. He was a very fastidious sort of chap, apart from his sessions on the booze. I'd ride in to the station on the Saturday to pick up the mail and I'd take a couple of packhorses in for a load of chaff and the groceries and the mail, and stay out on the Saturday and go back late Sunday afternoon. There's a whole week's lot of papers, so old Jeff had them hanging over a wire, and they're all in order from the earliest one – Saturday, Monday, Tuesday, Wednesday, Thursday, Friday. He only read one a night, the oldest one to the newest, but I'd go through more than one a night and when I'd finished them I always had to put them back in the right order. That was one of Jeff's foibles.[4]

The government deer cullers started work out at Golden Hills towards the end of 1938. Gordon could remember them all, a dozen working in pairs under Scotty McGregor, the head field officer, and a packman who took stores out and brought loads of skins in. They were all good keen men and 'one of them, Jock Munro – he was a Scotsman – wore kilts all the time he was on the station. There were cullers on Otupae and Mangaohane as well and that summer they took over 6,000 deer off the three stations. The following year the numbers dropped to 4,000, then 2,000.'

Gordon, soon promoted to foreman of the general hands and also station bookkeeper, moved into the homestead and lived with the family for the next eight years. When war broke out he waited for his call-up thinking 'when they want me they'll come and get me – I was on the waiting list to go into Waiouru'. As others left he took over their jobs: tractor driver, general chores, as well as doing the wages and petty accounts. They could not do without him so he agreed to Lawrence appealing his call-up. 'The appeal went through and I was there for the duration.'

Ray White did a season mustering on Ngamatea in 1941 before he was called up, and when he had had some time in the Army, on home service, Lawrence appealed for his release as an essential industry worker and he returned to Ngamatea. The shortage of musterers was a major problem. Incomplete musters meant increased numbers of double- and triple-deckers

The station hands' band on verandah of their whare.

– sheep carrying two or three years' worth of wool – out in the back country and they became harder and harder to muster. The head musterer in 1941 was Bernie Doole, who would do the mustering season from 1 October to the end of May. According to Ray White, 'he always had the same mate with him – Henry Isaacson. A full mustering team would have been six men, but I never saw a full mustering team. Lawrence must have been driven crazy trying to get musterers for a job as big as that.' When Ray came back from the Army, Bernie and Henry were still there, the only two remaining from the earlier gang, and a couple of others joined them: all had been turned down by the Army for flat feet or childhood asthma or weak hearts. They turned out to be fitter than a bunch of 100 or so soldiers and their officers camped at Golden Hills, in training for bush warfare in the Pacific. 'They got lost, broke arms and legs and shot each other; it was only their medics who gained any real training. We lost endless time trying to look after them and our boss got browned off with it all.'

The team went right through the season and took the wethers back to their particular blocks 'in the snow, in May. You could get snow on Ngamatea at any time of the year.' Ray had 'a couple of very suitable dogs' so Lawrence gave him the job of straggle mustering Mount Cameron in the winter of 1943. He based himself for the most part at the Log Cabin 'only about 10 miles back – the closest hut or whare of any of the camps'. It was built, probably in the late 1930s, by Atholl Parkinson who had only one arm yet he constructed it mostly by himself, with help and advice from Ted Isaacson, a brother of Henry, who had had experience in Canada. 'There were no nails, though there were plenty of good straight birch poles from the nearby bush and they were notched either end with an axe and fitted together at all four corners. But it had one bad disadvantage – the wetas used to come out of the rotten logs into your sleeping bags, and I tell you, if you got one of these big bush wetas crawling up your leg in the sleeping bag it wasn't very pleasant.'[5]

As agreed with Lawrence, Ray would be found 'in chaff, food, horses and everything' if he would camp out there and do a straggle muster on his own.

> That was a pretty tough proposition, but nobody wanted to winter at Ngamatea. It was just too hard a climate and nobody wanted to camp – none of the musterers that were capable of straggle mustering and bringing in roughies. And he put a proposition to me that at the end of the season he would give me 10/- a head for all the wild sheep and roughies – that's a station-bred sheep with the station earmark that's been missed by the original muster and become a hermit. There could be four or five years of wool on a roughie. They could only walk along very slowly dragging all this wool with them and Lawrence was always very conscious of wool. He loved this fine wool. I'd find groups of say 5 up to 20 – but the beauty of it was I'd be deer stalking and shooting and I'd wait till such time as I found a mob in a good locality, that it was possible for one man and his dogs to move. I had a wire yard at the Log Cabin where I used to work them round and educate them till I had enough to make it worth while going in to the station – say 50 to 100. It wasn't economical to bring in fewer. I could graze them and handle them in this yard for a couple of weeks perhaps, but I had to crutch them and clean them up a bit to get them across the Taruarau River. I didn't trace clip them, only the rear and underneath, and I would pack that wool in chaff bags.

Ray had his own hack, and one packhorse, and when he had enough sheep, wool and skins he would make the trip in to the station. Deer skins were worth a lot at the time because they were used to line the fuel tanks of aeroplanes: they 'would contract where there was a shrapnel or bullet hole and seal it to a certain extent and it made the difference for a bomber or any plane getting back to England'. He spent one winter shooting, partly in partnership with Lawrence. 'We both benefited – the Inland Patea was full of deer. The snow stopped me sometimes, but then I used to hunt the bush which wasn't nearly as bad as the open in the winter. I only did one season. I would never do it again – it was just too tough on your own. It would make all the difference if there were two of you.' But the next winter

Wild sheep.

Lawrence offered Ray a rabbiting block: he had done a lot of winter rabbiting in the South Island. 'So I either hunted wild sheep and deer skins in the winter, or took on a rabbiting block. In that cold hard climate skins were of top quality.' There was good money to be made between mustering seasons if you could brave the climate, 'and musterers' wages were quite good – £1 a day, as long as we stayed on the place'.

There would be some wild sheep in the mobs that Ray brought in from the Log Cabin, but keeping them together was not easy. 'They were like two different mobs: the roughies so slow moving, and the wild sheep so fast. You had to keep checking them with a heading dog that would lead.' Walter 'used to come occasionally, with a couple of dogs, and he'd go out as a sport and bring in a mob of wild sheep and we might be in at the station and we'd find them in the yards. Any man who could go out with a couple of dogs on his own and muster in wild sheep from the rough country, must have been a good man.' Out riding on the Hogget block one day, Lawrence and Winnie saw a wild ram on a rock outcrop high above them, bleating for its mates. Lawrence looked up at it in disgust: 'What the hell are you doing up there – blathering for Walter?'

Wild sheep, the descendants of earlier merino flocks, were found particularly on Mount Cameron. They were good eating – but really only good for dog tucker. They had no value and were just a nuisance: they spread ticks and lice and enticed other sheep away. 'The roughies with the

Roughies, or hermits.

station earmark were a different story. They didn't shed their wool like the wild sheep; they had a staple of 14 or 15 inches, and you can imagine the worth of those sheep to the station.' They would not be returned to the back country but would be kept in closer to prevent them becoming hermits again – and taking a mob with them.

> When I was there we decided that we would let a certain number of station ewes go on Mount Cameron hoping they would associate with the rams and we would be able to muster them in, but sadly that didn't work out. The wild rams just ran over the top of our heading dogs. We couldn't hold them at all, and they took the station sheep with them as well, so the boss lost his sheep. We did get them at a later date – we had to make a special muster to do that. But we got them back and we didn't worry about wild sheep or anything else – just getting the station ewes back. We used to shoot the wild sheep when they got too thick; we tried to shoot them out altogether because they weren't valuable. You'd see bare patches of skin with no wool at all, and bits of wool dropping off them.

Only once in Ray White's four years on Ngamatea did they ever have time to do a normal straggle muster to get sheep missed in the main muster. They straggled 'most of the Crown lands' and got 5,080 sheep, not, Ray felt, because they were bad musterers, but simply because there were so few of them and they had only from October to Christmas, when the shearers arrived, to muster all the back country into the Swamp block – the 18,000-acre 'holding paddock' that itself took a full gang to muster. They just had to keep going, and if they missed sheep they hoped they would have time to go back and straggle later on. At Golden Hills, about halfway in to the Swamp from the back country, there was a small netted holding paddock where they could overnight sheep before taking them on down the Taruarau Valley. In the 1940s there was a permanent government rabbiter at Golden Hills: 'It was a fairly central camp – easy country – almost a holiday camp for them there'. Then there was the huge shingle slide on the Manson, south-east of Golden Hills. 'I reckon we must have climbed 1,000 or 1,500 feet up there. When you got on the top it was easier going, but it was a terrific climb. Mustering off we used to go straight down, into the Manson Creek, and then a bad heavy zigzag going up to Surveyor's Rock, before the Log Cabin.' The Manson was walking country. 'We had no option – it was good country, warm country, did the stock well, but it was very rough.' Apparently someone – maybe R. T. Warren – had taken pocketfuls of grass-seed out there and in Ray's time you could still find cocksfoot and clover in sheltered places.

Lawrence would go out with the musterers, but 'he was a devil to muster with because he was forever trying to get a few roughies in because they paid him more than anything else'. They would meet him at the end of the day 'plodding along with 20 or 30 roughies in front of him with the station mark on them, and a grin from ear to ear. But he was a darned nuisance because we had that time-limit on us, so Bernie, our head musterer, was always at him.' For all that, though, Lawrence 'was a good man, he had a good heart'. Ray never knew him to sack anybody, and one time a couple of lads, just youngsters, rode in to Ngamatea on a big half-draught horse. 'No one knew where they'd come from, and all they had was a sheepskin and a surcingle, and the two of them rode there together and they looked half-starved.' Ray and the others took pity on them and sent them down to the cookhouse. They were brothers, Ash and Ron Dunstan, and when they had had a meal they were sent up to the house to see the boss.

> They came back all smiles. They both had got a job. Lawrence just for the goodness of his heart had taken them on – though we were always wanting men, but they were only kids. Ash was fifteen and I think Ron was only thirteen. They had a few dogs with them, and Ash had got a job as a paddock shepherd, and Ron had joined the rouseabouts. Ron did very well in later years. After he left Ngamatea he became a policeman – he was about 6 feet 2 inches and a good stamp of a lad he grew into. I think he went up the coast

– Gisborne way, to make his start. He was quite well thought of. Ash chucked in the dogs and came down country here fencing.

Lawrence could also be very uncompromising. Ray heard Gordon Mattson tell him that his dogs were worrying sheep. 'They were outside the gate, by the house and he came out with a shotgun and said to Gordon: "Which ones?" – "That one, and that one, and that one." And he took the gun and Bang, bang, bang – this is true – I heard the shots – and he said, "They'll bbloody well wworry no more."'[6]

In 1942, when Noel Roberts had just turned seventeen, he left Feilding Agricultural High School and went back to Ngamatea to work for his Uncle Lawrence. He started as cowboy, milking the house-cows, mowing the lawns, chopping the wood and filling the wood boxes, then became a general hand. The crawler tractors Ngamatea had at the time were his thing and a lot of agricultural work was done: oats for the packhorses and hacks, chou moellier[7] and swedes for the ewes kept on the front country. Noel lived in the house, and teamed up with Gordon Mattson, whom he had known since childhood. It was Gordon's idea to put in the Strip straight out to the main road from the junction where the station road forked – the left-hand fork heading towards Hastings and the other towards Taihape, and both of them clay. It took three of them to do the job: Gordon, Noel and the tractor driver, Dave Tibbles. 'We ploughed it both ways into the middle, then just graded it with an old horse-drawn road-grader towed behind the tractor. Then it was rolled and sown down in grass.' Now there were three entrances instead of two, 'but you could sail out there on the Strip in wet weather, no trouble',[8] and it was used until a metal road was put in in the early 1970s. Trucks could come in on the Strip, but not trucks with trailers. There were no trees between the road and the station in those days – but there were fourteen gates.

Extra hands in the house meant more work for Winnie. She had little help – at the most a married woman in the cottage at times to give a hand with housework. Bill Wells and his eighteen-year-old bride Evelyn were in the cottage for a year from July 1951. Bill was cowman and Evelyn helped in the house and the kitchen, under Winnie's tuition. After a year the roadman's job came up and Winnie said, 'Yes, you take it, Bill. We need a good man on the road.' So they moved into the roadman's house 'just a mile along the road from Timahanga, up on a bit of a rise on the left' and were there until January 1955 when the Oppatts took over. 'You just had to go along and make sure the water tables were clear, do the culverts when they blocked, cut a bit of tutu. You weren't overworked. You got ninepence a mile for running the Model A on the roadman's job, on your 12 miles – from the middle of the bridge at Kuripapango to the top of the Taruarau.'[9]

The Wells were Ngamatea's nearest neighbours, on the station phone line and included in Christmas festivities and various station activities.

Winnie's brother and sister-in-law, Ted and Joyce Craven, were in the cottage on and off. Ted was foreman and looked after the machinery, supervised the oat stacking and hay-making, and often took over the bookkeeping; Joyce helped with the cooking. That was a rare treat for Winnie. It was also a rare event for there to be only Lawrence, Winnie and the children at the table, as Jack explained:

> The bookkeeper ate with us, sometimes there was a stock agent, the vet, the wool classer, or a mechanic from the tractor firm in Marton. The married couple normally ate in the kitchen; he would be the tractor or truck driver or general hand, or sometimes the cowman or cowman-gardener. The cowboy and them ate at the cookhouse. At the peak, when everyone was there, they'd milk seven or so cows. Mum used to make a lot of it into butter. No fridges in those days. The milk supplied the cookhouse, but probably not the shearers – they probably brought powdered milk, because you might have three cows in milk, or six – they couldn't be sure. When we were hay-making there were extra people. Top dressers too.

The old house was a challenge, like everything else about Ngamatea. The best thing about it, according to several people, was the Aga stove: it never went out and it heated the house and the water. There was no other heating besides the fire in the front room and one in the cottage out the back. The house did not consume so much firewood, but the cookhouse, with its big double-oven Orion stoves, did. The brick-lined baker's oven also used a lot of fuel: you raked the fire out and put the bread in. There was no power – the hydro scheme was not put in until the early 1950s – and every day all the lamps had to be cleaned and filled. This was usually Winnie's job.

> For seventeen years a daily task was to pump kerosene and fill the lamps. The kero came in 40-gallon drums, the lamps mostly the well-known Miller with the round wick which always seemed to be in need of trimming, and the Queen Mary, a glass wall-lamp with tin reflector; also a similar type in tin, the Bulldog. Naturally we bought our lamp glasses by the dozen and our matches by the gross. They came in big wooden cases. With the advent of incandescent lamps our fireside evenings were a pleasure. The Aladdins were tricky; one got the mantle all brightly burning when a draught or a puff of wind ruined the whole thing, and if the door was left open from the outside a draft came in and the mantle burned black. Very soon everything in the room was festooned in soot. Then came the Coleman lights. Their hissing noise was

their disadvantage, but a great light. Then we decided, or perhaps the wool prices decided for us, on electricity. Lawrence would never entertain the idea of a small plant for the homestead; we would wait until we could have a plant big enough to supply Ngamatea.

However, while that was in progress we did have a small automatic plant at the house – just the homestead, the cottage, and the dear old gardener's hut. Being a temporary unit the engine was housed in a little back room of the cottage and the first light started the engine, and the last off stopped it. Aunty and Uncle were living in the cottage, so we tried to give them quietness, but poor Aunty, the gardener read until late and had a habit of switching his light on at any time of the night. The engine sprang to life, Aunty jumped out of bed with a yell. It was only in the next room and the whole cottage vibrated like a ship in a rough sea when the propeller was out of the water.[10]

The station telephone was another challenge. The house was connected to Taihape exchange by a one-wire earth-return telephone line strung on posts over miles of rough country. It always came down in the snow and someone would have to plod along the line until they could find and repair the break. It was a party line, with nine other subscribers – all the properties along the road as far as the Rangitikei River. Timahanga, Pohokura and the roadman's house had no outside line but were connected to the Ngamatea homestead, and Winnie or the bookkeeper would take all their calls and repeat them to the outside world. One roadman's wife had a speech impediment that made taking down her stores lists something of a guessing game. Another roadman had a daughter aged about 20.

She was courting some bloke who lived in town and Mum was the go-between for these two on the phone where they did most of their courting. 'Mrs Roberts, would you ring my boyfriend and give him a message?' And the boyfriend would ask Mum to ring the girl and give her a message. Mum thought she was broadminded, but she was a lot more broadminded at the end of those telephone conversations than she was before.[11]

Any of the station staff who really needed to make a phone call had to go to the one and only phone at the homestead. In those days there might be an hour or two to wait for a toll call to be connected, but they were always made welcome and would sit in the front room with the family and have a cup of tea.

Jack and Margaret were good children, busy with their ponies and involved with all that went on. Their great delight was to ride with their father, on his horse. One of Jack's earliest memories is of his father coming in from the muster. He and Margaret would race out to the gate to meet him.

Eunice Joll on Miracle, Margaret on King Billy, Jack on Oracle and Gordon Mattson on Blue Belle.

There was a bit of a rail there with a flat board on the top and we used to scramble up and wait for Dad. In those days he rode that great big mare Rosie, and she had a canter like a rocking horse. She could canter about as slow as you could walk. Dad would come in and we'd shoot up on the rail and he would pick us off, one at a time and sit us up in front of him and we'd canter round the corner by the old vegetable garden and round the back where there was a trough, and he'd give Rosie a drink there and canter back to the rail, and then the next one would have a turn.

I had my own first pony when I was about six or seven. I'd started to learn to ride on an old pony I think was left over from the Roberts boys – but I think he died about that time and there was only old King Billy left, and I learned to ride on King Billy. I used to ride him up and down on the big front lawn at Ngamatea, learn to rise to the trot up and down the lawn. Those rides would be my first memories of me and Missy together too. We were virtually learning to ride together. We rode with Eunice, and with Mum and Dad, quite a bit. Mum was a mad keen horsewoman.

Eunice Joll was their governess – their teacher, they called her – and she followed her cousin Eileen Joll, Jack's first teacher, who had stayed just one year. Eunice had started at Otupae at the beginning of 1945, teaching the four Lyall children, and she soon met Winnie – Mrs Roberts to her – during one of her visits to the Lyalls. Help with supervising Jack's Correspondence School work was just what Winnie needed, so Eunice wrote to sixteen-year-old Eileen and suggested she come up to Ngamatea. Eileen had no idea where Ngamatea was or how long it would take to get there, but in March 1945 she set out on a Mills Transport truck as far as Waiwhare, where she was met by Winnie, who had driven down to meet her and Mrs Hatch, the new house-cook.[12] It was all quite an experience for a girl who never been away from home before. Jack continued the story: 'After a few weeks Eileen came to Mum and said, "Mrs Roberts, I've got a confession to make – I've fallen in love with your nephew, Noel." And Mum said, "Don't worry about it, Eileen, you'll get over it." And she's still married to him! We went to their fiftieth wedding anniversary a few years back.'

Eunice had heard about the Otupae job through a neighbour and it appealed to her 'romantic mind'. Not keen 'to be locked up in an office', she could visualise the mountains, the hills, her own horse. 'I got excited but my parents weren't too pleased at the thought of a seventeen-year-old going way back there on her own. Mrs Lyall had sent a message to my mother that she mustn't worry about the boy situation – all the young men had gone to war.' Seeing that she was pretty determined, her parents suggested she go and see how she liked the feel of it. She left by sheeptruck at four in the morning.

> All those hours on the road. I loved it. And the Lyalls were very nice people to work for – and soon Eileen was at Ngamatea. They thought they'd better keep us girls happy and see we could visit each other. So in very short time Mrs Roberts came over to visit Mrs Lyall and brought Eileen, and who should be the driver but Gordon [Mattson]. We just shook hands – I didn't take much notice of the bloke – I'd heard he and Noel were nice fellows. I was going to teach Eileen to ride – I'd give her her first lesson on the boss's big old cob. I took her out for the day, and then they said 'Eunice you must come over to Ngamatea for a weekend with Eileen' and a few weeks later I rode over – 12 miles, 4 miles up the road and 8 or so shortcut through the tussock. And the solitude and the quietness. It was something that will stay with me for always.

Every three months Gordon would take Eileen and Eunice into Taihape to do their shopping. As Eunice recalled, her pay 'went a long long way. I bought a beautiful outfit with my first cheque – woollen suit £4, blouse, hat, gloves – all well within the £13 I had'.

> Later on Gordon used to bring me down to the Bay for the term holidays – he had stores to pick up for the station. We always dressed up to go to town on a Friday night in Hastings. People didn't go casually dressed – they used

to dress up for Friday night shopping. We'd end up the evening by going to a restaurant and we always ordered the same thing. We got a really good deal: a flounder, chips and two eggs. It was a plateful – 2/6, freshly caught on the Hawke's Bay coast, and a change from mutton. Then we'd wend our way back – a long way down to Otupae to drop me off again. It was a very enjoyable year there. I rode much of the station in my time off, even on my own if there was no one to come with me.

Towards the end of that year, when I knew Eileen was going to leave, it was arranged that I would take her place at Ngamatea and at the beginning of 1946 I moved.

Eunice enjoyed Ngamatea, the huge rides, often with Winnie, and living as one of the family in the house. The boys looked after her like 'a little sister', but they joked and teased her and started teasing Gordon as well: they knew they were already close friends. 'But the joke was that Mrs Roberts told Mrs Lyall it was hard to keep girls up there; if they were over twenty-one Gordon would be after them, and if they were under twenty-one Noel would be after them.' Before the end of the year Eunice left Ngamatea to nurse her grandmother, unaware that Gordon meant to leave when she did. 'We were good chums and all that but I had no idea that I was going to unsettle him to that extent.' They were married less than a year later – she was nineteen – and they visited Ngamatea on their honeymoon.[13] According to Jack, when Winnie lost her second governess in two years she said, 'The next person I get is going to be a man, because the next governess who comes might take off with Lawrence!'

Eileen had had just one pupil to teach; Eunice had two, but their routine was much the same:

> go through their lessons with them, clarify anything and supervise their work and tick it off when I was satisfied with it. And every month or whenever, the papers had to be sent to Wellington and every three months or so someone would come up – like an inspector, and really check out that it was being done correctly. The children didn't have the diversions other kids had, and when they left correspondence and went to boarding school they did very well – they had a really good grounding. Margaret and Jack stayed on correspondence all their primary years.

The girls were both teacher and nursemaid – they bathed the children and read them stories in the evening – but as part of the family they also gave a hand with other duties wherever they could. With so many in the house – Noel and Gordon, a governess and Nanny Craven – Winnie had other help at the time: the cook-housekeeper, Mrs Hatch. Lawrence was out mustering when she arrived: 'he walked into the house with all his whiskers and dirty clothes, and she ordered him out!'[14] Eileen listened sympathetically to the shepherds 'and anyone else who came with their girlfriend troubles. I was only sixteen, but I was more mature for my age. I helped them with their bleeding hearts. I didn't have to write letters for anybody or anything like that – they'd just come along and have a little chat with me, to help solve some of their problems.'

They seldom had other playmates but Jack and Margaret were perfectly happy with their own company. Margaret was 'a dear little girl', according to Eileen. She had a favourite blanket, a cuddle-rug she called Blankie. One day Winnie found it cut into four squares. 'Margaret, what have you done to Blankie?' 'Oh Mummy, I thought I'd do that so it would last me till I'm 21.' But she soon grew out of cuddle-rugs. Jack was Margaret's model: what he did she wanted to do. She refused to wear dresses and Winnie could not get her into 'anything nice and girlyfied' unless they were going to town, and then there were ructions because she had to wear something different from Jack. Eileen 'actually smocked her a frock – pale blue Vyella – and Winnie told me she'd had that for years . . . but Margaret wouldn't wear it'. The two children were always outside in their 'work' clothes, busy with such tasks as feeding a pet fawn with a bottle. 'We raised one or two,' said Jack. 'One went down to the Wellwoods, old neighbours of Mum's family, and when it grew up it became a bit of a nuisance. When kids were cycling past to go to school, it used to run alongside them for company and the parents got a bit upset thinking it might knock the kids off their bikes, and they had to destroy it.'

If Jack was Margaret's model, Lawrence was Jack's. Lawrence, ever the wool man, had a habit of picking up odd staples blowing around the woolshed or yards and putting them in his hip pocket. When Jack was only about five and Margaret three, Jack picked up the habit and Margaret imitated Jack. 'But not only that, when the musterers came in from mustering the 18,000-acre holding paddock and saw the two kiddies with a staple of wool in their hip pocket and the boss himself with a staple of wool, they had to be in the swim too. So we had five or six fully grown men and everybody wasn't fashionably dressed unless they had a staple of wool in their hip pocket.'[15]

The house was surrounded by a massive hedge – lawsoniana perhaps – which made a very effective shelter belt: 'it could be a howling gale on that side, and dead calm this side'. On the lawn in front of the house there was a big tree which the kids liked to climb.

> One day Missy came into the kitchen and said to Mum, 'Jack wants you', and Mum was busy and said, 'If Jack wants me, he knows where to find me.' And

Jack and Margaret with their pet fawn and Joe, the cat.

Missy hung round and hung round, and Mum said, 'Do you want a drink?' and Missy had a drink of milk, and then she had a biscuit and then she said to Mum again, 'Jack really wants you', and Mum said, 'If Jack really wants me, he can come and get me.' And Missy said, 'He can't.' 'Why not?' 'Because he's hanging by one leg from the top of that tree.' Then there was instant panic and Mum came racing out and I was hanging 5 or 6 feet from the top of the tree with a rope round my ankle. I'd climbed up and slipped and the rope luckily had wrapped round one ankle, otherwise I'd have come straight down, so Mum had to climb the tree and take my weight and unwrap the rope from around my leg.

Sometimes the cousins – the Roberts twins, Barry and David, or Anne and Bruce Craven, also twins – or sometimes all of them together, came up for holidays; and occasionally the family would go down to stay with their friends the Wellwoods at Irongate. Then there was Aunty Dossie, as she was known, the Roberts twins' mother, at Haumoana. Jack admitted that the beach was 'something new, but you soon got bored with it, because there wasn't all that much activity. And probably at that stage I couldn't swim, so the beach wasn't much use to me.' They learned to swim in the Taruarau or the old dam that fed the swim dip – that was the Ngamatea swimming hole. Christmas or early January was a great time because the shearing gang brought lots of children with them. 'We had the grass tennis court and with the long evenings the gang would come up after tea and play tennis till dark, and those who didn't want to play tennis would get cordial and cigarettes from the station store, so there was a lot of interaction. We were never isolated and lonely.'[16]

Winnie described shearing time as 'very sociable'. There would be ten shearers, ten shed hands, three pressers, an expert and a few sheepos, and 'ever so many children' – 40 or 50 people in all, and there was always a dance in the woolshed. The shearers' quarters held them all comfortably. 'In fact during the war the government authorities wrote and asked, in the case of evacuation, how many we could accommodate. We did say 50, anyway, because we had two sheds on the place, two lots of shearers' quarters, and I said the only stipulation was that our shepherds had asked for a girls' school to be evacuated to Ngamatea!'[17]

When their war service ended some of the men returned to the station: Harold Quilter came back as head musterer in October 1944. In June 1943 Winnie got an unexpected phone call from George Everett. 'Hello Miss Craven, it's George. Please can I have my old job back?' 'George, I'm Mrs Roberts now and I have two children.' 'That's all right, Miss Craven. Can I have my old job back?' He got it back, and a second rabbiter, Gus McMahon, usually known as Mac, joined him in April 1945. George then spent most of his time at Timahanga and Pohokura, and across on the Te Koau block; Mac stayed up on Ngamatea. Before Jack went to high school he would sometimes go down and stay with Mac at his tent camp in White's block, and later, when Winnie took a trip to England and Jack had just left boarding school, 'Mac came into the house to do the cooking and housekeeping and look after Dad and me. He used to be a boxer, and he fancied himself as that. If there were any scraps or stoushes Mac would be somewhere in the middle of it.'

When Gordon Mattson left towards the end of 1946 his place was taken by D'Arcy Fernandez; Winnie got her male teacher. He had worked for several years at a big accounting firm in Hastings, then ended up at Mills Transport, which had the mail run up to Waiwhare and did all the carting for Ngamatea as well as all sorts of odd jobs in town for Lawrence. They knew he was looking for someone to do the station books and teach the children, and D'Arcy was ideal: he knew the back country because he had tramped all over it with the Rover Scouts from Hastings, he hated being cooped up in an office and loved the outdoor life, and he was a keen hunter and a great shot. He stayed at Ngamatea for eleven or twelve years and came back for holidays or to help out for another decade after that. 'I was 21 when I went up there. I consider it the highlight of my life really, being there – and I still go back to visit or fish. I think the appeal was the vastness

Timahanga country from Otupae range.

of it, the tussock and the wildness and the remoteness. I'd spent time in the Chathams, where I had relatives – similar sort of remoteness.'[18]

D'Arcy had no sooner got to Ngamatea than Margaret, almost six, decided to show him around the place. 'We went for a ride out to the road and coming back we were cantering along, not far from the homestead, and she fell off. She complained a bit, but it didn't seem anything too serious to me so I put her back on her horse. When we got home she was complaining to her mother about her arm, so they rushed her into Taihape hospital. They suspected it was broken, and it was, but it didn't dawn on me.'

They started their school year on the verandah, where Winnie had taught Johnny Roberts, but after a while Winnie had the ploughman's hut made into a schoolroom. It was moved behind the house, the wheels came off and three desks were put in. It doubled as D'Arcy's office and remained the station office when it ceased to be a schoolroom.

> I'd never supervised correspondence lessons before, but they were good kids and Winnie had broken them in good and proper, and then they had the Joll girls. And they were very obedient. It was just if they struck any difficulty. I'd show them what they had to do for the day, and when they'd done that they were free to go. Margaret knew what she was doing – she was a bright girl. Jack wasn't as quick, but very thorough, like his father. I'd sit with them if I

Jack, D'Arcy and Margaret – with a broken arm from a fall from her pony.

was doing book work but if there wasn't any I'd set them up and go out and do something else – there was plenty to do. They'd sit there and do their lessons nicely – more or less. It was no problem keeping them to their school work, and once it was done they could take off.

The accounts were not such a big job for D'Arcy.

I only did the prime entries, the accountant in Wanganui did all the secondary, and the balance sheets and everything, but the store kept me busy. We were completely self-contained up there with the store. We used to get big orders from De Pellichet McLeod's, at wholesale rates. The shepherds could buy stores from us if they wanted a weekend away somewhere. And the shearers – it was cheaper up there because we never knew the prices! The minute the shearers arrived first place they'd go to was the store to see what we had. We had cigarettes, toiletries, some clothing. If the boys wanted something special we could order it from Taihape – the chemist or Farmers, wherever, and get it sent out by the mail and it was all recorded in the wages book.

It was Winnie who made up the order for the stores. They came in huge quantities from Hastings, by the truckload at a time – stores for the house, three cookhouses and the station store.

D'Arcy, Jack and Margaret on the steps of the Ngamatea schoolroom.

Access was difficult and the stores didn't come in very often. I think flour kept better then than it does now – it would go mildewed now. We got a ton of flour and half a ton of sugar – everything in big quantities like that. Two tons of potatoes, although we had a gardener who stayed 20 years with us, and we never were short of vegetables while we had Colin – ever. He was just marvellous. And we used to get our tea in huge quantities, and I think it was a unique way of getting tea. I would ring Hastings to Mills the carriers, and Jim Mills was a radio ham. He used to have friends all over the world that he talked to. He would be talking to Ceylon probably within an hour or two of when I had told him we were getting short of tea. We got our tea from a firm called Brown and Co and they had a ham with them who was a friend of Jim Mills and the very first ship that arrived there on its way to New Zealand had ten chests of tea put aboard for us. They came to Napier and Mills the carriers picked it up and it came on up to us.[19]

As well as flour and sugar, Winnie's store order included:

tea in 100 lb chests, cost in 1933 1/10 lb, cost rising steadily; golden syrup by the case, in 2 lb tins, also 6 lb tins; jam in 6 lb tins for cook houses, 2 lb tins for camps and mustering; butter in 50 lb boxes in winter, and in summer the station cows helped – the cowboy churned the cream in an Alfa Laval churn twice a week, and I washed and did it up in 1 lbs, wrapped in butter paper which we bought by the ream – years of this made me speedy – 25 lbs in 20 mins, and accurately guessed; blue peas, rice, split peas, lima beans – all by the sack; sago and tapioca in 56 lb bags; candles by the case – 10d per packet (Prices), 50 pkts to case; dates, sultanas, currants in wooden boxes – 4 lb; condensed milk (10d tin); matches in big wooden boxes by the gross; all brands of popular (at the time) tobacco, cigarettes and plug tobacco – the cheapest 9d a 2 oz plug, dark Havelock, Golden Eagle, etc; all stationery, e.g. writing pads, envelopes, stamps, inks, pens and pencils.

The only clothing she ordered was socks, sometimes £80 worth, which came direct from mills at Dunedin. Winnie also ordered anything the men wanted from Taihape. 'An invoice copy was sent to me and the cost taken off wages at end of month.' Winnie shopped for the men in town too. 'I bought their shoes, their boots, their pyjamas, their shirts, and I knew just what they wanted. Well, they were there for so long I just knew what suited them.'[20]

D'Arcy lived in the house with the family – 'Winnie was just like a mother to me' – and he would go out in the evening and maybe get two or three deer. At breakfast Lawrence would ask how he had got on.

Oh, I got three deer. 'Well, the packhorses are all down there doing nothing. Go and get one and bring it in for the dogs.' What a waste of good venison! We ate some venison – we had our share, and I always used to bring the tongues back for Winnie – she'd pickle them. She had a big stone jar and she'd toss them in and when she'd got enough she'd sort of compress them. Back steaks are all right – anything that was edible we'd eat, but the dogs got a lot. You get tired of venison. But you never get tired of mutton. And they had halfbred sheep, which is beautiful mutton.

I had the skins – that was my perk. I used to hang them up in the woolshed. I made just about as much out of deer skins as I did out of wages. We were getting 7/6 a pound, dried, for a skin – they might weigh 4 or 5 pounds. And we used to save the tails, the pizzles and the velvet and send them down to Harry Wong in Wellington. The antlers were worth getting. I hunted from the first – I've always been a shooter, and a tramper.

Jack remembered the rifle D'Arcy used – 'a .22 Hornet, which is just an overgrown .22. He used to shoot most of his deer through the neck or the head and hardly any of his skins had a bullet hole in them. He kept a tally of the deer he shot and Mum often asked him how many he had now, and he would just give a quiet grin and say "a few". That was about as much as they could get out of him.'

A regular visitor at Ngamatea in those days was Natalie McLeod, who had first been taken up there on holiday by her friend and fellow secondary teacher Olive Mays. Olive's boyfriend used to shoot on the station. Once Nat saw Ngamatea she was smitten: 'she spent all her holidays there,' said Jack, 'and once spent a year there while she wrote one of her books, supposedly helping Mum', but riding for much of her time, turning up at the nearer huts for a visit in the middle of the day, plaguing the life out of the boys. She was very outspoken, full of pranks – and she had her eye on D'Arcy, the quietest and gentlest person on the place. They married years later. Ngamatea was becoming quite a matrimonial bureau, but Jack and Margaret were away at secondary school by then so at least Winnie did not lose another teacher in the marriage stakes.

In the decade since the big scrub-cutting gangs had been working at Timahanga and Pohokura the scrub had come back. Lawrence tackled the problem by letting contracts for blocks, 'a couple of hundred pounds a block, and they'd get two or three people; one year they had a group of Aussies scrub-cutting down there'. In November 1948 a contract was let for the Spiral block and a year later Jimmy's block was cut. Four or five two-man teams worked down there but 'after the scrub was cut the land was never fenced or subdivided and the scrub just came back again, year after year; it's got to be fenced and stocked'.[21] The next move was to use machines to crush the scrub, and Lawrence called on his nephew, Noel.

> We disced our trial block from scrub – what we called the Starvation block, just on the right-hand side as you go into Timahanga. The trial went well – you can see the result today. I think that's one block that didn't revert, that stayed in grass – a block of some hundreds of acres. Then we moved down to Pohokura and did a lot of giant discing down there: the Bush block, the Lakes, the Sawpits. We did a bit of development on Ngamatea on the Oat Paddock Hill – we giant disced a block in there. But other than that we didn't do much on Ngamatea at all – mainly all down at Pohokura.

In the mid-1950s the big machinery was called in, and another great character, Gordon Grant, was introduced to the place and was there for

Noel and Johnny Roberts with the bulldozer and truck in Jimmy's creek.

the next ten or twelve years. Gordon worked for Tom London, a Hawke's Bay contractor, and arrived up at Pohokura a couple of weeks before Tom.

> I was to unload the tractors at Kuripapango – they couldn't be trucked over Gentle Annie – we had to drive them over. The big tractors we could only get as far as Willowford then you had to drive them all the way to Pohokura. This first tractor was only a small one. Mr Roberts was to meet me, and I arrived and all I could see was all this scrub and I only had this small tractor and I thought this thing can't live in there. I thought there was going to be 60 or 100 acres to do up there, short scrub. I said, 'Where do I start Mr Roberts?' and he said, 'Wwhat do you mmean where do you start? I'll tell you wwhere to ffinish.' So I got in touch with Tom and told him what the job looked like and he went off and bought a big tractor – an HD9, twice as big as the one we arrived there with. We had this big machine, with super giant discs and rollers, and Mr Roberts would sit there and watch us in wonderment. We kept a couple of the HD5s for lighter work – harrowing.

They started 'way down the back at Te Koau' and lived in tents until Tom built 'these two huts – only about 20 by 12, little wee wood stove – little kitchen and one bedroom'. Gordon got married a couple of years later and took his bride, Beryl, a trained nursing sister, back to Pohokura. 'She said she'd better not go and look first – but she liked it, and Mrs Roberts loved having a nurse on the property. Tom London's wife, Pat, and my wife were the only ones there. There was no one at Te Koau.' And there were no Land Rovers on the station, no transport, apart from tractor or horse, that could get them in and out of the valley. When they went to town they came out on a trailer towed by a crawler tractor. It took them over three hours to get up to Timahanga where their cars were parked – covered in mud in winter, covered in dust in summer. 'Much as I worshipped the ground Mr Roberts walked on, and still do, the only thing I never forgave him for was starting the breaking-in at the back, way down at Te Koau. Why he didn't start at Timahanga and work down the valley I'll never know.' The answer, of course, was that Te Koau and Pohokura were freehold, and Timahanga leasehold, but it did make the job difficult, and expensive: 'everything was back to front – we used to drag all the super and seed in – cost three times as much. But he was the boss and you did what you were told. He was a marvellous man though – the nicest man, without a doubt, that I ever had anything to do with, once you got to know him. I ran him back from town a couple of times, when his car broke down, and I had some great talks with him.'[22]

After a few years Gordon bought Tom London out and joined the two huts together, in an L-shape, so he had a three-bedroom house with a little verandah propped up on kanuka poles – his three children were all born while they were living at Pohokura. 'This was right down by the Sawpits, a mile, mile and a half beyond the cookhouse. They called it Londontown. Later we sledged the huts out to Jimmy's – just past his hut up to the foot of the Spiral. That Spiral was shocking – the old spiral. Steep. Terrifying! A long way to the bottom.'

The scrub at Te Koau was huge – mostly manuka and too thick to crush. They could crush red manuka up to 6 or 8 inches through.

> It used to go down very easily, and the denser it was, the easier to crush – it just goes down and breaks off at the ground, and it's like a carpet. Mr Roberts would say, 'Go over there and crush that lot', and you'd ask what the scrub was like and he'd say, 'It's so thick even the pigs have to come out to grunt.' There's a big scrub bar on the front of the tractor to break it down, and the super giant discs and the rollers behind would flatten it and you'd leave it for twelve months, more, and burn it – we had some fires all right. It's the scattered stuff that's hard on gear. That dense stuff that's just like a wall that you can't see through, that's beautiful, lovely!
>
> After Te Koau we went up above Jimmy's, back to the range, then to Fosters, up in behind Jimmy's, to the right. Back to Pohokura – the Bush paddock, then the Road block up at Timahanga and right down to before you drop down to the Spiral. There wasn't a block up there I didn't have

Gordon Grant negotiating the Spiral.

something to do with. And the thrill of my life – Grant's . . . . There's Top Grant's and Bottom Grant's. I'm very proud of that. A great thrill when he told me they were naming the paddock after me. That's where my ashes are going. Grant's is above the Taruarau.[23]

Lawrence had a road put in from Timahanga to Jimmy's in the late 1950s, but there was still no bridge at the Taruarau apart from the narrow swingbridge built by Drummond Fernie and Tau Wilson in the early 1930s as a sheep crossing. The road followed the old pack-track, which was 'quite steep in places, one in four. When the road was completed my father took the old Plymouth down there and to get back up he had to put the chains on and Mum wouldn't ride with him – she walked. She said with the chains on it was throwing the gravel right over the edge of the road into space. You really had to plant your foot to get up.'[24] This was Gordon's shocking Spiral.

In 1947 John Roberts came back to his old home as a shepherd and musterer. Lawrence gave all his brother's children employment at one time or another: both of the twins worked on Ngamatea. Barry was the truck driver for a year or two in the 1950s, and David worked with Noel when he first started doing the agricultural work at Pohokura. John was only eighteen or nineteen and he started down at Timahanga about crutching time. Ike Robin was the contractor. He shore for the Watherstons in the late

1920s and took over again, in early 1942, from Tau Wilson. In early 1951 he did his last shearing at Ngamatea; then his son Robbie took over. 'Ike was old – he was a New Zealand wrestler, and he weighed over 20 stone. He just used to sit in his chair, but he had complete control over his gang. They were a good gang. They did all the big Wairarapa stations – Lagoon Hill, all the Riddiford places, then they'd come up to Ngamatea.' Shearing was late up there and they always seemed to be shearing over Christmas. The shearing gang would not go home because 'they had a better time up there. All the station hands and the shearers and odd bods, we'd go down to the shearers' quarters and have Christmas dinner – the family, everyone – the whole station would go down. The homestead and two station cooks – or three – would all prepare a bit and everyone would be seated at big trestle tables.'[25]

When John Roberts started down there at the beginning of May, Ken Belt was head shepherd at Timahanga, but he left at the end of June to get married.

> I'd never had any responsibility. There were six or seven men there – fencers, packmen, shepherds, a cook there and one at Pohokura. I couldn't sleep with this responsibility – what was I going to do? I came to the conclusion I just had to get into it – I was in charge. But I said to the boss one day, 'Are you going to get a manager here?' and he said, 'No, you carry on.' That's how I was there two and a half years but I told Uncle Lawrence I'd like a season mustering. I got one, and enjoyed it very much.

Walter would come up to the property once or twice a year to see how things were going.

> He'd ride out somewhere – he had a great saying: 'Hi hi, Hi hi'. You'd hear him coming – 'Hi hi, Hi hi'. He'd bring a few rough sheep that hadn't been shorn. He'd 'just happen to see them' so he'd capture them and bring them back. We'd find them in the yards, shear them, and they'd be dog tucker. I never went off the place for seven months. 'You take over boy', and you took over.

As John recalled, 'Ngamatea was a world in itself. You never looked at your watch – if you had one. You started at daybreak and when it was too dark you stopped. You always had a billy on your saddle with tea and sugar and if you wanted a feed and you had the time you stopped. The job was there and you did it. Time was no object.' There was no set day off – the boys took time when they needed it, 'but most of the time we'd be chasing pigs or deer'. The Hawke's Bay Show was an annual event 'and if we'd wanted a weekend off we could have had it, no problem, but we just didn't'.[26]

Lawrence preparing to blade shear his precious Bibbenluke stud rams.

Lawrence struggled to increase the stock numbers. They grew slowly in the early years, but between January 1948 and January 1967 the main shearing tallies varied between 21,000 and 28,000, with similar numbers at the beginning and the end of the period and the low point in 1962. It is difficult to make comparisons because some of the records refer to Owhaoko and some only to Ngamatea – or Ngamatia. The change in spelling from Ngamatia, as many still pronounce it, to Ngamatea was gradual: it vacillated for some years in the early 1940s before it settled, but as far as Lawrence was concerned the name for the combined properties was Owhaoko, and that was the brand he used for his wool. It was a return to the name the Studholmes used for the property, although they pronounced it 'Ohaok' – and still do. There are old maps and wool brands around with various spellings: Ohauko, Ohauku, Owhaoku, Owhaoko.

On the lower sheltered country at Pohokura Lawrence ran a small English Leicester stud of about 150 breeding ewes and replacements, and at Ngamatea his precious merinos, for many years the only merino stud in the North Island.

> Dad used to buy merino ewes from down south – from Jopps at Alexandra and Gibsons near Tarras, and also from around Marlborough. One year when he brought the ewes up from Marlborough it was the biggest consignment of ewes that had ever been flown up till then. They were flown to Napier and

Winnie Roberts with a Bibbenluke ram at the Ngamatea woolshed.

trucked up from there. He bought some rams from the Wright family of the Bibbenluke Stud in New South Wales – Dad and I went over to the Wright property, north-west of Sydney. He also bought rams from a stud down near Alexandra. So he had this merino stud and also the English Leicesters, and the merino rams went across English Leicester ewes to get the halfbreds. The Ngamatea flock was halfbred. All the Fernie places ran fine-woolled sheep.

The Wrights had three boys and one girl, Mary. She was an air hostess, and she and another hostess, Margot Vincent, came to Ngamatea towards the end of 1957 for a holiday. Margot stayed on and did the books – and left in April 1959 with Ken Sinclair, one of the musterers, and they were married in Sydney soon afterwards![27]

From 1953 to 1967 the Crown continually had an eye on the Ngamatea leasehold blocks. Most were Maori lease, but the Crown was particularly concerned about part of Owhaoko D7b, an 8,574-acre strip between Peter's and the Taruarau, the only part of the Gift Block leased by Ngamatea; it had been proclaimed Crown land in 1918. In April 1953 the

Department of Lands was thinking that 'the best approach to protection of the catchment against burning and overgrazing would be the acquisition by the Crown of the station to the south, and the Maori land between the station and the Gift Block'.

It just happened that 51,588 acres of Owhaoko D7a and D7b, in negotiation since 1954, was two years later 'being acquired by Mr L. H. Roberts', and as this 'ruled out any development scheme by the Crown' they 'decided against purchase'. The district valuer from Wanganui had spent five days inspecting the back country in March 1956 and the report he wrote might have helped them make up their minds.

> The whole area comprises steep broken country chiefly in birch bush and scrub; but with areas of bare wind-eroded surfaces, and steep rock slides, and smaller areas and pockets of tussock grazing. The bulk of the country lies at the 3,500 ft level, with areas up to 4,600 ft, and the main streams cut down to 2,000 ft. Access is by 6 ½ miles of private unmetalled road to Ngamatea Station, while musterers' tracks circle most of the country, giving access to musterers' huts. There is a considerable area of birch bush. . . . of little or no commercial value because of its location and access, but it is available for fencing, firing and framework of huts. The stock on this country from Ngamatea get most of their grazing on tussock on the Crown land.[28]

The best idea now seemed 'to dispose of the balance of Gift Block – about 18,360 acres – to the best advantage'. Fernie Brothers and Roberts, the Crown thought, would be the logical purchasers especially as Lawrence Roberts had also applied to buy the 8,574 acres they were already leasing. Admitting that nothing 'has been done with the land by the Crown' and that Mr Roberts 'would like to see the block re-vested in the Maori owners so he can buy it from them', the Director of Crown Lands recommended the block 'be sold to Mr Roberts, proceeds to be paid to the Maori owners, subject to the consent of the Tuwharetoa Trust Board and approval of the Minister of Maori Affairs and the Minister of Forests'.[29] The sale did not eventuate. The trust board wanted the Gift Blocks – never used, as intended, for the settlement of returned soldiers – given back to the Maori owners; and the Department of Lands was finally starting to recognise that its grandiose notion of developing the back country was unrealisable.

When the question came up again a decade later, in August 1967, Lands and Survey noted that it had done an aerial inspection in 1962 but a ground inspection was needed and had not yet been made. 'We haven't pressed for it because of the low value of the land. It is not suitable for development. The northern portion should be shut up for soil conservation purposes, and the southern portion has value only for extensive grazing by Ngamatea station . . . whose stock graze over Maori and Crown land – including part of the Gift blocks – for which there is no tenure, as well as freehold and Crown lease. The total area grazed could be 150,000 ac.' A week later, however, the Director of Crown Lands wrote to the Secretary of Maori Affairs: 'the Crown notes the successful development on Otupae and Mangaohane and is interested in Ngamatea Crown lands for a development scheme'.[30]

Lawrence's purchase of Owhaoko D7a and 7b made a huge difference to the station. He had added about 45,000 acres to the 5,000 acres of freehold they had formerly held, and now owned White's, the Tikitiki, the Lake block and Peter's – out to the present back boundary. There was nothing now to prevent development – except Uncle Walter's conservative ways.

Lawrence Roberts was a great judge of men; he was honest and he was fair but he could be tough, especially if someone was trying to put one over him or was not pulling his weight. Ray Birdsall, his head musterer in the mid-1950s, knew Lawrence well.

> He'd have no hesitation in arriving to pay somebody off and he'd look around and he'd say 'And who else wants to go?' And you'd be short handed before he said that too, you'd be struggling to get the work done. . . . When a chap decided he'd had enough and he was going to leave and he didn't have a vehicle he'd ask could he be taken out to the mail truck. 'Hobnail Express' he'd be told and he'd have to walk 6 miles out – more – 6 to the road, then the trip along to the mail-box – a couple of miles there.
>
> He was a character – but he was loyal – he was loyal to his good men. When you get a lot of men together there's always rows about something. One evening when I was head musterer there must have been a bit of a row, a fight or something and somebody said to Mr Roberts, 'There was a scrap last night'. Next morning when I started work he turned to me and said 'I hear there was a scrap last night. Don't let them get the first hit in. You gotta drop them first.' I never forgot that![31]

Lawrence was particularly wise in the ways of young musterers. As Jack told the story, he would have one of them lined up for a 5 a.m. start at shearing.

> His job was to bring the sheep forward from the back of the shed because they had the big lean-to pens at the Ngamatea shed – the covered pens, and it was

a helluva job because the sheep had to come up from the pens, go through the shed, turn round at the top and then start coming back down the race to go into the catching pens. Peter Cameron was the one designated to be up at the shed that morning, and five o'clock came and he wasn't there, so Dad stalked off down to the musterers' quarters and Peter's still sound asleep in bed – they've had a bit of a party the night before. He stirred Peter up and stepped out the door again to go back to the shed, and Peter thought he'd sneak a couple of extra minutes in his warm bed and the other blokes are all barely stirring, and he said to nobody in particular, 'Has that old bastard gone yet?' And Dad was waiting just outside the door, and he stepped back in and said, 'No, he hasn't. I'm still here.'[32]

He had a great way of disciplining the musterers too, teaching them their place in the scheme of things – but only if he judged them worthy. A likely lad needing a bit of a lesson would be called to work in the yards with Lawrence and a small mob of his merino rams in a big pen. 'Catch that one,' he would say, and the young fellow would go after this ram and grab it by its great handle-bar horns and wrestle it to the edge of the pen. The boss would have a bit of a look at it and say, 'Now trim its feet, and put it in this pen over here.' When that struggle was over, he would say, 'Catch that one' – and it would be inspected and its feet trimmed and put in another pen. This continued until all the rams had been caught and had their feet nicely trimmed. Then there would be instructions to take this two or three to that paddock and this one or two somewhere else, until there was one pen left, with most of the big old rams in it. 'Take them down to the killing house,' he would say, 'and kill them for dog-tucker.'

The musterers knew they could learn from Lawrence too, just watching him sort the ewes before they put the ram out. 'He had a long broom handle, a bag of crushed raddle tied to the end of it, and he just dobbed them on the head or on the shoulder as they came through the race. That represented a certain type of wool, and that sheep went to a certain type of ram – halfbred, merino, or whatever. He was a great bloke, and he was very good with his wool, there's no doubt about that.'[33]

Lawrence was a hard man to get to know but when he spoke it was to the point – and often humorous. And he could speak in public when the occasion arose, and he could not avoid it. Apparently he and Ted Craven 'would unwind at the RSA and get up to all sorts of mischief down there', and he was friendly with Neil Dorreen, the Wellwoods' farm manager, 'and played a game or two of social cricket for the Flaxmere Cricket Club'.[34] Winnie was everybody's friend, a mother figure to everyone, loved by all: 'what a wonderful woman she was, a fantastic woman'.

Winnie and Lawrence Roberts at Ngamatea.

> If anything went wrong, if you cut your finger or did something like that the boss would say in his gruff way, 'Go up and see Mrs Roberts.' Peter Cameron got a bee sting working in the sheepyards one day and within minutes he had swelled up. Old Lawrence sent him up to the house to Mrs Roberts and she painted him with Reckitt's Blue. She said, 'Strip off Peter . . . I said, Strip off, Peter – I've got to paint you with Reckitt's Blue.' That was what we used to do in those days.[35]

Another of the boys remembered Winnie as 'a lovely person – very friendly, very warm, and I think she was a great comfort to a lot of the young fellows who've gone through Ngamatea over the years, and perhaps had different backgrounds. We didn't see a lot of her, but we always knew she was there, and she always had a good word when we did see her.'[36]

The boys probably did not know how highly she thought of them: 'Over the years we had a great bunch of boys. We were proud of them when they went to functions – they inclined to keep together as a group. At balls the lot of them were invited, the station bought the tickets and the cost came out of wages at month's end, and the Ball Committee supplied partners for our boys. The blind dates were a great attraction for both parties!'[37]

Winnie was very hospitable: she liked visitors and entertaining people of all nationalities. The Swedish ambassador and his wife came for the weekend with Lena, a Swedish girl Winnie knew – not that Winnie realised this pleasant unassuming man was the ambassador. 'What do you do at the embassy?' she asked, and he had to admit to being more than just another Swedish civil servant. They quietly left a perfect Swedish crystal vase at the end of their visit. Winnie always called it her tear-drop vase: it was just that shape. Americans loved to come to the station. One couple from California, Harold and Mary-Lou Wickersham, heard about Ngamatea through one of Jack and Margaret's Correspondence School supervisors and wrote to ask if they could come as paying guests. Winnie replied that she did not have paying guests, just guests, and that they should come. They did, and were entranced by the place. Harold always had a camera round his neck and, according to Margaret, he was always where the action was: if he was not in the middle of the yards he would bob up out of the tussock as the musterers came towards him and cause havoc with the mob.[38]

There were some visitors Lawrence was not comfortable with and when they were expected he would find urgent business out the back somewhere. 'He had two grey horses, lovely hacks, one he'd pack and one he'd ride. When Winnie had organised for some of these visitors to come and stay a couple of nights, you'd see Lawrence get his horses and pack them up and ride away to the Tikitik, the nearest hut – just to get away from them I suppose, because it wasn't his thing.'[39] Riding in from the back one day, he met a Land Rover out beyond the Woolwash.

> Some government experts or something. They didn't know who he was and they stopped and asked how far they could go with their vehicle. He said 'You're there' and just kept riding. A bit later they were back at the house wanting to know if they could get a tractor to tow them out. When they saw him they thought, 'That's the bloke we saw before', and he said, 'I told you you were there.'[40]

The right sort of visitor was another matter altogether. After Paul Thomas left Ngamatea in 1960 he went overseas for a couple of years and soon after he got back Winnie phoned and invited him to bring his father down from Wairoa for Easter.

> They were dipping down at Jimmy's while we were there and Lawrence wanted to go down early next morning. I was to go down later and take Mrs Roberts and the lunch box, but I suggested to Dad he might like to go early with Lawrence to open the gates – there must have been 20-odd gates I suppose. I told him he'd have to make the conversation: 'The old boy isn't a very outgoing sort of a bloke and sometimes he'll just grunt at you'. Dad didn't think that would be any problem at all – and when we got down to

Lawrence Roberts.

Jimmy's they were already the best of mates, talking and laughing together and I said 'How did you get on coming down?' 'Oh, great. It was pretty much silence as you'd suggested for the first two or three gates and then we got on to the war.' They were both First World War men, serving about the same time in France, quite close to each other – in the trenches, of course. From that point on they got on like a house on fire and you couldn't stop either of them talking. They really enjoyed each other's company.⁴¹

Lawrence and Winnie were respected by all who knew them, on the station and off. Tony Batley recalled Lawrence as 'a man of few words – taciturn, reticent, almost shy. But we had a lot of respect for him. They used to have an annual ball here at Moawhango, and every year he and Mrs Roberts would come in, and we always thought it was wonderful. It was a trek in those days – they didn't have a metal road, or anything.'

The musterers particularly felt that working for Lawrence was the experience of a lifetime.

> He was such a team man. That's one of the greatest things we learned – there was no room for individuals. And he would read you like a book. If there were any misfits he'd take them out. There was one musterer – first season for him, and he'd been at the head shepherd more than once and told him he deserved more money than the others because his dogs were better than theirs. The head shepherd had to pass this on to Mr Roberts when we got back in

to the station and a bit later the boss phoned down to the head shepherd and said, 'Ththat boy with the ggood dogs. Ttell him I'm going to Hastings in the morning and he can ppack his gear and I'll take him with me, hhim and his ggood dogs.' End of story – and we went on one short for a while. That's how the man worked, no mucking around. If someone was disrupting the team, he was better off out of it, no matter how good he and his dogs were. He taught us lessons for life. [42]

Chapter Four

The Muster

It's not a farm; it's a bloody province.

The muster of the wethers from the back country was the crucial event of the year. Everything depended on a good team of musterers and a clean muster. The season began on 1 October each year and lasted about eight months: the end would come when Lawrence announced, usually without warning, 'the season finishes today'. A full mustering team was seven musterers – 'the boys' – and a packman-cook, with eight hacks, a team of a dozen or more packhorses, and swags of dogs – up to seven per musterer. It was a great spectacle to see them riding across the Woolwash bridge and heading out into the tussock. Even Air Force planes, exercising in the area, would drop down to have a closer look. There was simply nothing like it in the rest of the country.

The success of the team depended on the skill of the head musterer in the man-management stakes. Lawrence chose his well and had faith in them; and he knew how to delegate: 'he used to come out the back – a couple of seasons he came out with us, but he took all the orders from the head shepherd. Never hesitated. Camped out in the huts with us – he was one of the gang. Never interfered with the head shepherd's orders. He knew his men; he knew the duds too.'[1]

Musterers were distinguished from shepherds in that they were paid by the day and employed for the season – although sometimes the terms were used interchangeably. They all arrived, or were due, on 1 October and were paid that day for travelling and for every day they were on the station. In the mid-1950s musterers got £2 10s, the second musterer £2 12s 6d and the head musterer £3, with double time for public holidays but not Sundays. If they went off the station for Christmas or any other day, they were not paid.

The season started with a horse muster. Having spent the winter turned out in the Gully block, all the horses had to be brought in and shod, and enough chaff packed out the back to maintain them through the muster. Thousands of wethers could find enough feed out there but horses in work certainly could not. In the off-season the musterers went down country,

Packhorses crossing the Woolwash bridge.

many of them to Hawke's Bay, to do a lambing beat, but one or two who had stayed over the winter to work as shepherds around the station and down at Pohokura would be on hand. Some years one of the musterers would come back a few days before the start of the season to help get things organised. Bruce Atchison did this several times: 'We'd get the horses up from their winter block and Forby Minto, the blacksmith from Otane used to come up and shoe them. Basically we had three horses each – two good ones and an old one. Then there could be sixteen or eighteen packhorses and they had to be shod too.'

The head musterer would decide who had which horses. 'Maybe one of your horses was a young one that had just been broken in. Of course they hadn't been ridden – it was a rodeo the first morning. The horses would be put in the big stable with the alleyway up the middle – you'd each be given your stalls and you'd saddle these horses up and you got on them inside the stable. Then they opened the front door and it was all on.'[2] If a musterer arrived late he got the horses nobody else had wanted. A year or two after he had left the station Dave Wedd was asked to come back when the season was under way and they were short-staffed, and he brought a mate with him: Arthur McRae was an East Coast drover, and he was new to Ngamatea.

> Most of the staff were there when Dave and I arrived – they'd been there a while and had the horses sorted. We had the left-over horses. I got a most unusual looking one – a two-colour thing, cream and grey, sort of splotchy – attractive to the eye, but nobody took it. I said 'What's wrong with that

The pack team loaded with chaff at the rear of the stables.

one?' – 'Oh, no . . . nothing.' I asked what they called it and I thought they said Ghost. Oh, that's it – now I know – it bucks. Rodeo name. You know – the Ghost, the Widow Maker.

We got up next morning and put it in the yard. I'd said to Dave 'I'm getting up early in the morning, I'm not putting on an exhibition for the boys. I'll get it under control first'. They still had the big old stables then. They were huge – wonderful. . . . I caught this horse and saddled up, got on – and it was good as gold. Of course. Under a roof – horses won't buck under a roof. I said 'Righto – open the doors'. And it just marched out – a beautiful horse. Not a sign of bucking – one of the best movers on the station. Nobody knew I suppose. I think the name must have put them off.[3]

While the shoeing was under way the packman-cook and one of the musterers would pack chaff out to Golden Hills, the central point for the main muster. They would take enough to see them through the first part of the muster – from Golden Hills up the Ngaruroro Valley to the Boyd and Mangamingi huts, then back to Golden Hills. This might be the packman-cook's first introduction to packhorses; they would go out one day and back the next and leave the chaff at Golden Hills for use there and for distribution to the other huts.

Used to have a few problems, the horses would be a bit frisky and buck the bloody things off; get a sore back or a sore wither, and you'd have to leave one or two behind. Same with the hacks – some of them used to buck, having the winter off. We knew which ones bucked and you'd all line up and old Lawrence Roberts had the great happy knack of knowing – he used to keep an eye on it – and sure as the Lord made little apples, if someone was going to have to ride a bucking horse next morning the old boy would be over at the woolshed, peering through the door to see the performance.

When everything was good and the horses had settled down and they were fit, it would be four, or say five hours, to Golden Hills. When they were coming home they were a bit faster. They were like ducks, they weren't tied together, they just followed along – they knew the pack-tracks and away they went.

If the weather was good in late October we'd be up between the Ngaruroro Valley and the Kaimanawa for up to the thick end of three weeks out mustering, but the packman-cook used to resupply meantime from Golden Hills. When we'd done the Ngaruroro Valley we'd come back to Golden Hills and a couple of the boys would go in with the packman, take all our spare hacks, get new stores and bring new horses back, and then we'd go to Kaimanawa.[4]

All the stores would be packed out the first day of the muster, along with the boys' swags.

We used to have a limit of 40 pounds each. That way a horse could take five swags – two on either side and one on top. And some of the new musterers, they used to take – well, if they had their way, they'd have taken the kitchen sink. You'd take the clothes you stood in, you had two other sets of clothes, your sleeping bag and other bits and pieces – gumboots, spare pair of boots. Some of the horses had special loads, like Trooper – and Kitty. She used to take the big pack-boxes with the bread, and we had those that used to take our swags.[5]

You'd arrive at the 4- or 500-acre Golden Hills holding paddock, someone would go and kill a couple of dog tuckers, someone else would hang up a mutton. They were pretty lean at the best of times, and it would go straight into the camp oven and by tea time it was beautifully cooked. It would go in just about kicking – it was warm – and it would come out like chicken. It really was – lean two-tooth halfbred wethers – they were just beautiful; you wouldn't believe how good. Well, we might have been pretty hungry. Bruce Atchison was a good cook – they were all pretty good.[6]

In the late 1940s there was a fairly high turnover of musterers. After the restrictions of wartime people were feeling free to move around a bit and several took the chance to do a mustering season at Ngamatea. Dixie McCarthy worked for many years for the Batleys, down the road at Moawhango, but he took a sort of leave-of-absence in 1945 and arrived at Ngamatea for the start of the season. 'I really enjoyed it. After you come away from there you think what a great experience – I wouldn't have missed that for anything. I would have liked to have sent my son up there. You never forget it.'[7]

Trevor Topliss was there for the 1946–7 season. He was from the South Island and had mustered for three or four years in the Awatere and Inward Kaikoura.

> Ngamatea was similar – tussock country – but different sort of tussock: the South Island has smaller white tussock, not red like Ngamatea; and it's not quite as steep as the South Island – though patches at Ngamatea were pretty rough; very steep next to the Rangitikei and over next to the Kaweka Range – the Manson country.
>
> We didn't camp out much mustering down south – we came back to the cookhouse most times, or maybe a different place altogether, on a different block – but not huts. Ngamatea was totally different in that respect. In Canterbury the places are divided by big rivers: we used to boundary ride our horses up each river, and get off the horses and start walking – climb up. The horses used to work their way home if you were lucky. Sometimes they didn't come home and you had to go and look for them.
>
> And of course we brought the sheep down for the winter, whereas at Ngamatea they put them out the back for the winter. In the south there were some straight merinos, some halfbreds. I'd worked with both so I was used to fine-woolled sheep when I got to Ngamatea. Different sort of sheep to handle from Romneys. Can't push halfbreds and merinos too hard – you've got to stand the dogs off. My dogs were used to them.
>
> I enjoyed the season – quite an experience. Ngamatea is probably the only place in the North Island anything like the South Island.[8]

Don Hammond knew the Roberts family from way back; they were 'uncle and auntie' to his wife Alison whose people, the Dorreens, were neighbours of the Cravens and the Wellwoods. Don had heard a lot about Ngamatea and wanted to experience it for himself. 'The mustering team that season, 1949–50, was Ash Watt, John Roberts, Sonny Barrett, young Tom Harker, Jack Cornwall, Hilton Cheesman and me. I came home and went shearing the next season. I would have liked to go back to Ngamatea, but it was the money side of it in those days. It was certainly a good experience. If I had my life over I'd do it again and I'd probably stay there longer.'[9]

A load of chaff at Golden Hills.

Bill Cummings went to Ngamatea in July 1945 and was there for four years – as head musterer from October 1947. 'I'd still have been there,' he said many years later, 'only I had to go home to the farm – my father took crook.' There were seven in the team in his first year: Terry Potham, head musterer, Ralph Atkins, second musterer, Hartrey Dampney, Jack Cornwall, Ivor Riddiford, Dixie McCarthy and Bill.

> You wanted two heading dogs on Ngamatea – it was too hard on one, especially straggle mustering. You know the mountains up there? – Mount Dowden on the right, and across the Mangamaire to Mount Donnelly – well, we always called it the Donnelly, but on the maps it's Makorako – it's the highest point on the place. And Mount Michael is pretty high. Sometimes we mustered the Dowden across onto the Donnelly, sometimes back to the Mangamingi, depending on the weather. And the Donnelly was mustered right away down towards Mount Michael, down the Mangamaire. Then you'd muster Mount Michael and bring them across the river, up what we call Tin Kettle Spur, and the Zigzag, sort of opposite the Michael, across the river – you come up the face to the top of the Kaimanawa. And Forks Spur is just down from there. The main fork of the Ngaruroro comes out of that big bush between Ngamatea and Taupo – heads up towards the Poronui track. We called that the Right Hand Fork. The other fork goes up towards the Mangamingi.

> Ashley Watt was rabbiting, culling, and ended up mustering at Ngamatea – on my team. Then he was head shepherd after I left. The head musterer is responsible for everything that happens out the back. You've got to make some decisions on whether to muster that day, or if the weather's too crook, or if you're out there and you have to call it off. I would as soon be just an ordinary musterer.[10]

Don McLean did the 1948–9 season – Bill Cummings's last as head musterer – with Ash Watt, Ian Sinclair, Hilton Cheesman, Ralph Atkins and Ted Gilbert in the team.

> Seven of us went there, but then it went down to five and back up to six – not a consistent crew all the time. Tom Harker joined us at one part of it – Young Tom. I didn't know Old Tom. And Brian O'Connor . . . he broke in some horses that year and handled them very well, but wasn't so keen on riding – we took them over. . . . Brian also took on packman-cook for a period, but the main packman-cook was Ian Moore ex-navy, great cook and a great fellow, excellent guy. And D'Arcy was there.

Don McLean remembered the five conversations he had with Lawrence while he was at Ngamatea.

> The first was when I arrived up there. I went up on Lumsden's Cascade beer truck, with the dogs – they were underneath. They had special boxes for them – single boxes, under the tray of the truck. They took shepherds and dogs up quite often. I was going from the Hastings side, don't remember where I got on, but I had been working at Glenross. Someone else was already in front so I was under the tarpaulin with the empty beer barrels. It was snowing, and when we got up to the Ngamatea road there was a bloke with a horse for me, to take the dogs, and Lawrence had his car, to take my swag, and he had to pick up the mail, and he said, 'Yyou McLean? I'm Rroberts. Not exactly tro-opical, is it.' And that was the end of that conversation. But when we worked drafting sheep at the yards, which we did quite a bit, he always gave us a 'Good morning'. We always greeted each other, but it didn't go further than that. He was reticent, but he wasn't disagreeable at all – he just wasn't talkative.

In the seven months he spent at Ngamatea, Don left the station only once – to go with the boys to a rodeo in Raetihi. That night Lawrence rang the shepherds' quarters 'just to find out how we got on. Of course we hadn't done very well at all. It was just a break away from the station – several of us went, not the whole lot. We didn't advertise ourselves as the Ngamatea boys – we weren't kidding ourselves that we were cowboys at all. But he was interested, and he'd have been happy if we had done well.' According to Don, if Lawrence thought enough of you and wanted you back he'd ask if you were coming back next year. 'He asked me, so I had that honour, but I said I was going to the South Island. . . . There was no discussion – why don't you stay, or anything – it was just accepted. He probably had a word with Bill about those he might ask to stay on – and he knew Ash – he'd been around a bit.'[11]

Ashley Watt, who was from Wairoa, had spent some time out the back of Ngamatea as a government deer culler; Gordon Mattson met him at Golden Hills in 1939. He started at Ngamatea as a shepherd in the winter of 1947 and joined the mustering team at the start of the next season. Don McLean enjoyed working with him: 'He was good with dogs, had a useful team, was good to muster with, a good mate – and he was interested in native plants and had quite a knowledge of the environment generally. He and I saw blue duck on the Mangamaire – a pair. We took a detour, thinking we just might see them, and we did.' There were blue ducks on the Kaimanawa Stream too – Waingakia on the maps. Dave Wedd saw them much later, in the early 1960s: they would come right to the door of the Kaimanawa hut for scraps thrown out for them. There were several pairs on the stream, each in its own territory: 'We used to see a lot of blue ducks while we were stalking through the bush. They never flew – they just peep peep peeped away, quite cute things. They had hardly ever seen man – they'd be within a few feet of you.'

One thing Ash hated was wetas, and the musterers encountered plenty of them – they were in the firewood and in the framework of all the huts, which were just malthoid stretched over 'birch' saplings. 'You'd turn your torch on and see them jump round in the night. But before they went to bed they'd start talking wetas, and they'd stick something in the bottom of Ash's sleeping bag and he'd put his foot on it and they reckoned he'd hit the floor running, and he wouldn't come back in until they tipped his sleeping bag inside out and proved it wasn't a weta.'[12] Ash was at Ngamatea for five and a half years, and stayed on year-round for all but two winters. He was head musterer from May 1951, and when he left in September 1953, Lawrence wrote in the wages book 'a good man'.

On his 1950–1 team Ash had a young fellow called Chris Lethbridge, a 'bright boy' and a good boxer, but not quite the usual young musterer, having come from a private school background and a year or two of down-country shepherding. He had heard about Ngamatea from D'Arcy Fernandez; they had met in the Scouting movement and had both been to a world jamboree in France. Chris arrived in August and spent the next few months puttering around the station. D'Arcy kept an eye on him:

Pack team and musterers crossing the Swamp block on the way to Golden Hills.

He had trouble with his dogs for a start. We had a few lambs that had got mixed up and Lawrence wanted them taken over to a paddock behind the woolshed, and Chris was there doing the odd jobs. Lawrence told him to take these lambs and put them in such-and-such a paddock and next minute we see Chris with the wheelbarrow, and he'd tied the lambs' legs together and put them in the wheelbarrow and wheeled them – a load or two. Of course you'd need good dogs to move half a dozen young lambs in the right direction, but Lawrence stood there looking at him and said, 'That's it – I've seen everything now!'

Chris was supremely happy with his lot, though, and set off with the musterers when the season began, but he had been out the back only ten days or so when he sprained his ankle badly over on the Harkness and was off work for three weeks. He finished the season and fully intended to come back for another one but his life took a different turn and he went into the Anglican ministry instead. He had not finished his studies when he had a much worse accident, spent five years in hospital and finished up in a wheelchair. He recorded his season on Ngamatea in *Sunrise on the Hills*, which was published in 1971.

Ash's place as head musterer was taken by Jack Cornwall, but he had 'a crook ankle, and couldn't stand the walking, so he chucked it in after a season and a bit'. Jack had mustered there previously then had a break, and when he came back he brought a mate, Ivan Chapman; it was not uncommon for musterers to be recruited this way. Ivan, known to many as

Shorty, stayed for four seasons under the new head musterer, Ray Birdsall. This was the mid-1950s and Ray kept his mustering team together with few changes for several seasons.

Ray had started in October 1951. He was 20, a Hawke's Bay boy like a lot of the other musterers, but he had had no contact with Ngamatea and became interested when he saw a newspaper advertisement for musterers. When he mentioned to a shearing contractor that he was thinking of going to Ngamatea, the response was: 'Oh, don't go near that place – it's terrible. I can get you plenty of jobs on Otupae, Mangaohane, any of the big places up there, but don't go to Ngamatea – it's a terrible place.' But Ray answered the ad and Lawrence phoned to offer him a job.

> I had to have my own saddle, dogs – I'd been shepherding for four years by then. I got in my little Ford 10 sports car – whipped the back seat out. Had my dogs – five or six of them – saddle and everything in the back, and away I went. I'd never been up the road before. I arrived up there about three o'clock in the afternoon on the main road where the station road turned off – the Hastings end. I got about a mile and a half in and it started to snow. It just poured down – and there I was, stuck. I'd let the dogs out and they were racing around in the snow – first time they'd seen it. Luckily it was a wire-spoked-wheel car, and I had a rope so I wrapped it round the wheels, acting like chains, but no, I still couldn't get up the hill. It was where the intersection was between the old Taihape end and the Hastings end of the station road and I was stuck there, trying to work out what I was going to do. And some riders came along – three of them. One was Bruce Atchison and another was Nat McLeod – coming back from the Otupae block – and Bruce said he'd get somebody to come out and give me a tow.
>
> Eventually a chap called Dick Sturm arrived out. He was a wood-cutter and a casual hand and he had big chains on his car – a V8 – and he had a heck of a big long rope and he got way up on the tussock and towed me out – towed me right in the 5 or 6 miles to the station. It took ages – we arrived at nine o'clock at night – no tea or anything. But Bruce was the real hero of the thing. He was a great joker, Bruce, and he took me into the cookhouse and there was an old cook there. He was a really good cook – he'd been the cook in England for the Lord Mayor of London – but he was a real hophead. All the tea was gone, but he made me some nice sandwiches. Anyway I just settled down, then went over to the whare and we had a big fire going, and I tied my dogs up and fed them and everything was right – and that was my introduction to Ngamatea.
>
> Next morning Bruce took me up to the house to meet Mr Roberts and everything went on from there. We were the only two of the mustering gang that had arrived. Bruce had spent the winter there because he'd done the year

Pack team on the way to Kaimanawa hut from Golden Hills.

before, and Mr Roberts said the head musterer, Ash Watt, was arriving on the Overland Transport and we'd better go out and pick him up. So we got on the old Bedford truck, and away we went, and along comes the Overland truck with all these big empty Cascade barrels on. The beer was made in Taihape and it was carted across to Hawke's Bay and the barrels came back empty. And Ash's dogs running around on top of the barrels, and Ash is sitting inside the truck. This is the public transport. It was the first time I'd met Ash. He was a really good joker, Ash, a really good joker.[13]

Ngamatea was short of musterers that season so it was just Ash, Bruce and Ray and a packman-cook who headed off on the main muster. They were out at Mangamingi hut when Alan Kennett arrived; he'd mustered the previous season but had just got back from the Chathams. Alan was an enthusiastic country and western singer: 'he used to cart a guitar round to some of the huts with him, the handier ones. He was great on that guitar, he could nearly make it talk, and he could sing and yodel. He would sing in the hut at night, make up little ditties. He was a character.'[14]

Ray tried to get Alan interested in shooting, but he reckoned he could not shoot. He had borrowed the station .303 one day and popped off round after round without hitting anything. Then they discovered the barrel was split 'and what he was using for a sight was just a bit of twisted metal'. One day Alan and Ray saw three deer down by the Taruarau. 'They took off up the hill and I said, "Go on Alan, use my rifle. You'll get them." "I can't, Ray." "Yes you can – there's a telescopic sight." Reluctantly he sighted the gun. Bang, bang, bang – three deer. He was ecstatic, and became quite a shooter.'

Looking down the Mangamingi from the pack track between the Mangamingi and Mangamaire; Dave Wedd with Tom.

A couple of days after Alan turned up one of the government deer cullers arrived out with a new musterer, 'a joker called Jim something'. He walked into the hut fairly late in the afternoon when they were having a cup of tea. Ray looked him over:

> I thought – you've got the same sort of spurs as I've got – I'd left mine in my bedroom back at the quarters. Ash says 'Pick the bunk you want' – and he spreads his gear out and Ash looked and he thought – they look like the same blankets as I've got. And he'd taken everything out of our rooms, because he had no gear, and he'd brought it all out the back. But he proved to be absolutely useless and he threw it in after a few weeks.
>
> We went through for a long time then with just four of us. We went right through shearing like that then a block of land just at the bottom of the Blowhard – you know, the other side of the Annie – was cut up for Rehab and there were two shepherds on it – Morrie Mott and Nani Te Momo. They had their own horses and Mr Roberts must have arranged to take the chaps from there. Anyway we were just finishing shearing and these two chaps rode on down the station road and they became two of the musterers.[15]

Ray was made head musterer on 5 December 1954, the day after Jack Cornwall left. He knew Lawrence had his eye on him and he had decided that if he became head musterer the first job when they got out the back would be to repair the horse paddock fence. There had been problems and 'a bust-up' over fellows arriving out for the muster and finding next morning that all the horses had headed back to the station and they had to walk all the way back in to get them. When Ray left to get married at the end of the 1957 season, Lawrence wrote in the wages book: 'Best all round head musterer so far. Would be well worth trying as a manager.' It was a prescient remark. Several musterers who worked for Ray were full of praise for a man who led by example and was excellent to work for. Ivan Chapman thought he must have had 'a heart like a lion'.

> He could muster all day, and always took the biggest beat, the hardest beat. When you went out to work with Ray of a morning – you worked. You respected Ray. He had six or seven musterers under him all the time, and you were getting up at three, half-past three, every morning, morning after morning, and you were doing a reasonably good long day and then jokers would get scratchy after a while. Everyone had a different temperament you know – you had to treat them all a wee bit different, and he could manage them all. He made a great job of it. I reckon he was absolutely A1. But the moment that the work had finished and you were home he was just one of the boys. He was a bit of a larrikin you know, always up to something. At the end of the day he'd come home and be the first one to have a cup of tea – and away out deer stalking, and he'd deer stalk till dark. He shot hundreds of deer up there.[16]

Alex Lindsay worked on Mangatapiri, but would come up and do part of a season at Ngamatea when required. He was there with Ray, Bruce Atchison, Morrie Mott and Ivan Chapman.

> Ivan had the best team of dogs around – even better than Ray's – dogs that could do things on their own. We were sent to take a mob somewhere, he and I, and we had to go across this bit of swampy ground and he had a couple of dogs he just put up the side, give 'em a whistle, and they'd walk, one from either side, in behind the lead, then just take the lead away so that the rest of the sheep would follow. That sort of thing. He had one good old dog and if anything got away in the scrub, he'd send this dog in and sit down and roll smokes and wait. Eventually this dog would arrive back with the sheep. You wouldn't know where they'd run to in the scrub.
>
> Sometimes you'd go damn near all day without seeing the tail end of a sheep then at the end of the muster, there would be the mob. I was lucky – I was put between ones who'd been there before and they kept me in line. The old hands were always put on to the difficult spots and they knew where to be and what they were doing, and when to work their way round, or where to wait for the muster-off, to catch them, in case they drifted away somewhere else.[17]

When Ray left Ngamatea Morrie Mott took over as head musterer. He would go down to Hawke's Bay to do a lambing beat between mustering seasons and he knew he was going to be short-staffed for the 1957–8 season as Ivan and two others had left when Ray went. So Morrie looked around, listened to the stories that circulated around the Bay, chose his staff and arrived back with them. That was fine by Lawrence: he had faith in Morrie's judgement. One of those he recruited was Frank Brady, who described Morrie as 'a fantastic stockman, though not a good dog man – but much underrated – a fantastic little man'. Everyone liked Morrie and most of them had a story about him. Jack remembered an incident during the shearing at Timahanga when Lawrence and Morrie were walking over to the cookhouse from the shed. They had to walk past the old malthoid shepherds' hut and 'there was a bit of a stoush up going on between two shepherds over something, and Morrie was going to step in between them and stop it and Dad said, "Oh, leave them be, Morrie, they couldn't hurt each other." They heard him – and that was the end of the fight. It just stopped dead.'

On another occasion Morrie and Ivan were turning out sheep and had arrived at Golden Hills, wet through. They had just got to the hut and it was still raining. 'We didn't expect Mr Roberts to come out, but all of a sudden there he is in the hut doorway and Morrie's down on his hands and knees trying to get the fire going, blowing it – wet firewood – and Mr Roberts appeared and Morrie turns round and says, "Can you cook?" And Mr Roberts says, "There's cooks and cuckoos, and I'm a cuckoo."'[18]

Morrie had a bad accident in the most difficult possible place: he broke his leg the day they were mustering off the Manson. Bruce Atchison remembered it clearly:

> We had no hacks over there, it was too steep, it was all on foot. He had turned the sheep onto the Manson shingle slide to bring them down to the bottom and he slipped on a bare patch where the frost had lifted the vegetation. Jack Roberts was our packman at the time and he found Morrie sitting there when he came down. He had to leave him there, and when we got down the bottom and got the sheep mustered off he told us Morrie was sitting up there with a broken leg. Cobber was the quietest packhorse. We unloaded him and took him back up and got Morrie on the pack-saddle and brought him down to the river – the Manson Creek – where we were having our lunch and all the sheep were. We made some splints out of manuka. I think we brought him down without them and made the splints down at the river. Everybody used to wear high-topped boots – they came halfway up your shin and they were good for crossing creeks without getting wet feet. We put these splints down his boot and I had a big scarf and we wrapped that around the top above the boot

Dave Withers and Bruce Atchison on a walking beat on the Manson.

– no bones sticking out or anything like that. Then to get him out was the big problem. The sheep used to go straight up the hill but the packhorses went up through the scrub, sort or zigzagged around the side to get up onto the top. Jack was to bring him in, to the station. Everyone had their lunch, including Morrie and we got it all organised and away they went. Morrie reckoned the worst part of the whole trip was going through the scrub because every now and then his foot would get hooked in a scrub bush and twist it. They got up the top to the Rock Camp and Jack caught the horses – the hacks – we'd leave them over there when we mustered the Manson. Morrie went in on his horse and it was quite lateish that night by the time they got to the station. They stopped the night there and next day put him in the car and took him in to Hastings hospital. That's about two and a half hours. In this day and age it would be helicopters and all floating around.[19]

The leg Morrie Mott broke was plated from an earlier break at Matapiro, and the plate had come loose. Jack wondered 'how the hell he stood the pain – he didn't say a word about it', and Frank Brady reckoned Morrie was 'one of the toughest chaps I've ever worked with'. After the accident Lawrence wrote in the wages book: 'Left to be married. A good man, unlucky, had several accidents'. Morrie had a very stable mustering team – there was hardly a change in it for two or three seasons. But things were different after his time, as Frank recalled:

You dug a hole in the side of a sandy bank and that's where the dogs slept.

Guys would say, 'Oh to hell with that – we weren't going to live in dirt-floored huts' and that sort of thing. But that never entered our heads. We had these sack bunks, we had a broom made out of manuka, we'd sweep this dirt floor and get any dust and that out of it. Who expected a carpet floor out there? At the station we thought we were in a motel. We had a wooden floor and everything, fireplace in the middle of the thing, chip heater. But when they built the new whare, the new shepherds' quarters at the station, I heard that it had only been up a matter of months, and they were skinning a deer on the carpet. They were a totally different type of crowd that had moved in, a younger type. We loved it because we liked the environment, and that was the difference. That's why you were there – it was a challenge.[20]

The day would start early out the back – at two in the morning if the muster was to begin a fair way from the hut; the musterers had to be on their beats before the sheep had a chance to start heading back to familiar territory. 'You didn't know if it was a late night or an early morning. You were always out catching your horses by torchlight and feeding them and letting your dogs go. The dog kennels – well, you used to take a shovel and you dug a hole in the side of the bank – that's where the dogs slept: in a sandy bank.'[21] The packman-cook would be the first up, making porridge, chops, tea. When the boys had caught and fed their horses, they would have breakfast, let the dogs go and head out in the dark and cold.

THE MUSTER   103

Some would take bread, chops, a bit of cold meat for lunch; some had a 'hillbilly', a small water container with cup and folding handle, a bit like a thermos, but 'you could dangle it over a bit of a fire and boil up'. Ash Watt always carried one on the side of his saddle. Ray Birdsall, it was said, was not a fan of stopping to boil up; he would just have a drink of cold water and keep going. But there were times when the fog was down and no one could keep going, and then little fires would spring up here and there while they boiled up and toasted their stale bread. Sometimes they were lucky and there would be a spectacular sunrise, with Ruapehu clearly visible on the horizon. From the top of the Dowden – Mount Dowding – they could see Taupo. The Dowden really is the parting of the waters. From its northern and western slopes the streams run into the Tauranga–Taupo and eventually into the Waikato and from its southern and eastern slopes into the Mangamingi, and then the Ngaruroro. The Rangitikei headwaters rise a bit further down, on Makorako – the Donnelly.

The muster would be over by mid-afternoon and the boys would arrive back at the hut looking forward to a feed and thinking about an afternoon's shooting. But if they were moving between huts the head musterer would detail one or two to go forward and get some mutton and dog tucker. When they neared the hut they would put the dogs round a few sheep, any they could get hold of just as long as they got enough for them and the dogs, kill and gut them, then go in for a packhorse and pack them back to the hut. 'There was a steep face by the Manson hut. If the advance party got there quick enough, carrying a .22 rifle, he could shoot one or two off the big face and if he was lucky they'd roll down towards him.'[22]

Packmen-cooks were a varied breed – 'we had good ones and not so good ones and useless ones' – and when there was none at all the boys took turns at cooking. 'The head shepherd would get everything organised so you'd go with the rest, but you could get away early and get home and get the meal on, or stick the meat in the camp oven before we left and sort of part cook it and then wind her up again when you got home.' Not many people had had experience of cooking with a camp oven so the boys had to break them in. One fellow, an ex-wharfie from Auckland, saw an ad for a job as packman-cook and thought he would have a go. 'Didn't have an idea what it entailed – didn't even know what a bloody horse looked like, virtually. But actually he turned out quite good. Then we had another little fellow – he was running away from his missus – he was up there hiding, and he'd never had much to do with horses, but he took to it quite well too.'[23]

One of the great success stories at the job was Ian Moore, who was not long out of the Navy when the boys met him in Hawke's Bay. He was looking for a job and the head shepherd at the time, Terry Potham, asked him if he knew anything about horses.

He said, 'No, not a thing.' Terry said he wanted a packman-cook, and he said, 'What the hell's that?' And he turned out a great bloke. Marvellous patience with horses, and he stayed there for quite a few years. Great cook – he picked it up, no trouble at all. You've got to load all those packhorses on your own. It's a big job – except when we're at the station, loading the chaff; we gave him a hand there. When we're camped out the back, and we're away early mustering and he's got to shift camp that day, he does all that on his own.[24]

Ian was packman-cook for three seasons and a casual hand and wood-cutter on the station for another two after he had an accident that left him with a limp. He never lost his Navy ways of cleanliness and order, and years later in the mid-1960s when he phoned Ngamatea to see if he could get a job, Winnie told him to come up – that there would always be a job there for him. He cooked at Timahanga then, on and off over the next five years on working holidays.

The musterers' huts were all built more or less to the same pattern, yet each had some distinguishing feature. Golden Hills, the most used hut, was a little more roomy than the others and had a bit of a verandah along the front where they could pile firewood and hang wet oilskins. Just outside the door to the right of the hut was a big mound of wood chips. They cut all their wood there and they would leave a good stockpile for when they returned, but hunters and others would drop by and use it all so Bruce Atchison made a long pile of wood chips with boots sticking out one end, a hat the other and a big cross marked 'Here Lies The Last Bastard Who Pinched Our Firewood'.

The huts were all a simple oblong, about 14 by 12 feet, with a corrugated iron fireplace and chimney at one end, a door at the other and four bunks, two up and two down along each side. The bunks were just a chaff sack nailed across two beams, 'and you had to wriggle round a bit because sometimes you had the two seams right across your bum or in the middle of your back. If you were lucky you had a bit of a cupboard on the wall, maybe a table, but some of the huts, by the time you got out of bed you were virtually sitting at the table – if there was a table. Otherwise you sat outside, or on your bunk.' In many huts bunks served as beds, seats and living space; they would try to put a tall bloke and a short one in two adjacent bunks to give them a bit more leg room. In their first season musterers were given a top bunk. 'Then if you didn't shake all the candle grease, rat shit and everything out before you got in, you got abused because every time you moved it poured down on the poor sod underneath. You learned that one very quickly.'[25]

The huts were made of beech poles cut out of the nearby bush, with the roof and sides covered with malthoid. They had a dirt floor, a 'little wee window right up the top', and all were cold, draughty, dark and smoky.

THE MUSTER

Golden Hills hut.

But to the boys they were 'home'. Living cheek by jowl with seven others for weeks at a time in such a tiny space was a test of anyone's character. 'I reckon under those conditions you either hated a joker or you got to like him. You get in one of those little huts, and she's full of smoke, your eyes are running and you're coughing and spluttering, there's no room to move, no little sheds out the back to put your wet clothes to dry – they'd be all hanging round the fire trying to dry out, your meals off your knee – no dining table, and no sitting in an easy chair when you're finished.'[26] And as the season progressed everyone would get tired and scratchy, and it was only the head musterer's skill that held everything together. When the weather was bad the boys would look forward to a day off. Once they badgered Bruce until he said, 'Righto, we're not going to do anything today', and they settled down to enjoy their leisure. But they were all bored in no time and next morning, rain, snow or fog, they could not get out of the hut quickly enough, and Bruce heard no more grizzles. 'That was just the way those experienced men taught us younger ones. It was a wonderful thing.'[27]

The Log Cabin did not conform to the usual pattern of huts. It was beautifully built, but it was not big enough for the gang so they had to pitch a tent alongside, and after 20 or 30 years the moss between the logs had disintegrated and 'it leaked like a bloody sieve'. When Jack Roberts was packman-cook he and Frank Brady were in the hut with all the pack-boxes, gear and food, when it started to blow like fury and the rain drove through

Inside Boyd hut; Dave Withers, Dave Wedd, Frank Brady, Guy Brittan, Bruce Atchison, Gordon Wright.

the gaps between the logs, bringing all the wetas out of the woodwork. 'You could hear them plop onto your sleeping bag cover, then you'd hear scuffle scuffle scuffle and if you had a torch you'd shine it down your bag and here would be several wetas walking up towards you.' Dave Withers and Dave Wedd were at the Log Cabin towards the end of one season when the rain was torrential. 'There was no way we could camp in the bottom bunks because the water was rising above the mud floor. There was a bit of a table, and we had the tucker up on that. We camped in the top bunks and in the morning we jumped out, and hello, woop, down into the water, ankle deep.'[28] The Log Cabin was destroyed by fire in the early 1960s. The musterers came home and it had gone – and all their gear with it.

The reading matter they took out the back with them did not go far. Winnie would get a supply of 'little square western magazines' from a Hastings bookshop and the boys would take them to read on wet days – 'the sum total of our reading matter – apart from your diary, if you took one. Some did, some didn't. I'd take out five or six of these westerns, and they'd go round the hut.'[29] They would read anything and everything in the hut, including Weetbix packets and the labels on sauce bottles, as Frank Brady remembered.

> I went down to Timahanga one time with Mr Roberts. We were coming back over the top, through Ngamatea and I said, 'Who broke in all this land for

Log Cabin; Dave Withers and Frank Brady.

you? Who was the contractor?' He said, 'Aw, Bryant and May.' He never says much, so you didn't think any more about it, but one day I was out the back and we were snowed in. I was sitting there, rolling a smoke – and you read everything, from books on – and I was reading this matchbox, and I suddenly spotted Bryant and May – they made the matches – and I burst out laughing. Of course the others couldn't understand and I told them what old Lawrence had said.

Most of the huts were on the edge of the bush and near a creek, which was the bathroom and laundry – not that much washing of any kind was done during the muster, except perhaps for socks.

You had to look after your feet there because you were walking a lot. You daren't wear dirty old socks – it didn't do your feet any good. So you just put a rock in the toe of each sock and dropped them in the creek – not in the swift part – in a bit of a backwash. They'd be in there over night and next day you'd give them a bit of a rub. No soap, only a bit of Lux soap to wash your face with in the morning, that's all. You'd go out with your towel around your neck, and a torch. It would be dark, you'd go over to the creek and all those creeks were *cold* because they came through the bush, high bush. They were really cold, it would wake you up all right.

Wash day out the back.

> The toilet was a spade – you dug a hole, and that was your toilet for each hut. But one day at Golden Hills it was wet or something and we had nothing to do, so we built a toilet. There's a creek just down from the hut door – quite a fast-flowing creek, but only about 10 foot wide, and there was a big log across it – we used to walk across. So we put up a few rails, and we put a seat on it, and that was the toilet. We put some old chaff sacks around it for a bit of privacy and we stood back and admired it. 'Now that's very hygienic', we said. We were skiting about this toilet. But the moment anyone went and sat on the toilet, someone would sneak out round the back of the hut, get a big stone out of the stream and come up behind you and just flop it into the stream and all the water would splash up round you everywhere. You had to watch out and see where everyone was when you went to the toilet otherwise you got wet through.[30]

On one occasion when this happened the victim was 'standing there giving them their pedigree and next thing he happened to turn round and here was Dr Bathgate just on the other side of the creek with five or six birdwatching people who had been walking in to stay in the hut, and here's this poor devil standing there with his pants down round his ankles'.[31]

Newspaper was at a premium out the back. It served many purposes from protection for certain things in the pack-boxes, to wrapping lunches and lighting fires. The same bit of paper might wrap lunches for three or four days, then find some other unmentionable purpose. But it was in short supply 'so normally bits of grass or something like that was used for toilet paper'. The boys would take soap, toothbrush and tobacco, but no shaving gear; at the Kaimanawa hut one season they took a photo of Ray Birdsall shaving off his beard with a pair of shears – 'he reckoned it was starting to itch'.

> One time we'd been out at Golden Hills for quite a while and Morrie Mott thought he'd boil up some water in a kerosene tin over the open fire and have a decent wash. Bruce came in and said Morrie was round the back there stripped off having a sponge down with this warm water. So we thought, Right – and everyone grabs a billy and down to the creek and gets a billy of cold water, and it was *cold*. Old Stan Taylor, our packman-cook, was a bit slow and there was no billies left except the tea billy. We'd had a cup of tea and there's a couple of inches in the bottom of the billy. So all these boys tore round there and they threw the cold water at Morrie, and old Morrie is standing off abusing hell out of them and old Stanley comes round with this billy of tea and throws it over him and of course all the water runs off him and all the tea leaves stuck to him. He was like a spotted leopard, and he'd used up all his water and he's left with all this cold tea running down him.[32]

When they got back to the station it would take two days for everyone to get a bath. If they ran the hot water cold the cook would abuse them; the bathroom with its one tin bath was around the back of the cookhouse. 'Then someone painted the old bath with aluminium paint – and the first bugger that used it got aluminium all over his backside because the paint got soft and stuck to him. We didn't give it long enough to dry.'[33]

The Manson hut was the least like home. It leaked and the floor turned to mud; the boys wore their gumboots inside. There was a bit of bush nearby, but the hut was built on the side of a steep hill with the creek way down at the bottom and steps cut into the bank for access to the water supply. Even less washing than usual went on at the Manson, but if the packman-cook suspected anyone might be thinking of going down to the creek, he would thrust a couple of billies at them and ask them to bring water back for him. In an idle moment one wet day Bruce Atchison heated a poker in the fire and burnt a message into the beam above the fireplace: 'Was the bastard who built this hut man or camel?'[34] One time when it rained and the water 'poured in' Jack Cornwall got a three-cornered iron fence standard that was lying around and positioned it so that the water ran down into a billy he put in the centre of the hut. Next morning he had his own water supply, and a dry bunk. A less inventive fellow simply went without a trip down to the creek: 'he never cleaned his teeth or anything for four days'.[35]

At the Manson in the early 1950s; Dave Bower, Alan Kennett, Ash Watt, Bruce Atchison, Ray Birdsall, Jack Bindon (packman), Harry Bimler (deer-cullers' packman).

The Manson country was the worst to muster. The musterers were out there early in the season and the weather was often bad; they could be snowed in for days or they might have to wait for the Manson Creek to go down before they could cross it. Sometimes they were stuck out there until the food got low. 'It does test you, and you did pit yourself against others. That was all right, but there were many, many hours of misery out there when you were confined to camp. You got sick of playing cards, sore throats from smoking too much, wet clothes and that sort of thing. But you tend to forget about that – and the sun does come out eventually.'[36] They made their own fun and found humour in each other's predicaments.

> We were out at the Manson one terrible day. We arrived out there just on dark, raining like hell, had a job to get the fire going, but Ray in his wisdom thought that if he heated up a stone in the fire when we did get it going, he could warm up his feet with it in his sleeping bag. He gets his stone and rolls it up in paper, then I think he dragged a sock over it to keep it all together, and put it in the bottom of his sleeping bag. We were all in our bunks and 'Jove,' he says, 'this is lovely in here, nice and warm.' He's moving round a bit and a while later moving round a bit more, twisting and kicking. Then, 'By Joves, I think this is getting warmer.' All of a sudden he jumped out of his sleeping bag and threw this thing out on the ground and when it got out in the air, the paper burst into flame. We all laughed our heads off of course.[37]

The Manson country was the worst to muster.

The muster-off of the Manson was 'always a great occasion – a big mob, all coming single or double file down the narrow ridge to the creek crossing, up into the Rock Camp area – down steeply and up steeply. But from years of mustering the blokes knew where to go, what to do, how to do it. Each had worked under another head shepherd, and the information was passed on down the line'. The Rock Camp was a cold place. It was what they called a fly camp: there was no hut, just a fireplace with a bit of corrugated iron to shield the camp ovens, a tent for the musterers and a fly to cover the horse feed. 'We'd do the Hogget, the Cameron, the Log Cabin area, right down towards Golden Hills, muster out that end, then go to Rock Camp for a couple of nights, and on to the Manson.'[38]

River crossings could be a problem, especially getting over the Ngaruroro from the Harkness to the Boyd. The timing had to be right; it was deep enough at the best of times, but snow melt or heavy rain would cause a hold-up.

> You had to start them – merinos and halfbreds would get in a big ring and you'd have to get in there and split them up. We'd get to the river, and we'd have to have a decoy. One man would use those little dog chains, or rope and a dog collar, and use one sheep as a decoy – ride across and tow it with you, and you might have to do that two or three times to get them started. It wasn't an easy job – it was tough. You'd be pushing and shoving – and we had to be careful they didn't all pile up and smother. Then the river might pick them up,

> and a few of them would get washed down into another area, and you'd need a man down there, to get a dog round to pick them up. I always remember the effort, coming out of the Harkness and across the river. Same sort of thing at the Taruarau, but less likelihood of a smother. It was a better crossing place – the rapids were shallower, stonier.[39]

Gold Creek was different again. The mob would not walk into the creek but they could be forced to jump from a bank opposite a flat piece where they could walk out, but again care was needed to see they did not pile in on top of each other.

The Mangamingi was the most distant hut – beyond the Boyd, 'right up there at the top end' in the headwaters of the Ngaruroro, past the Right Hand Fork. It was a cold hut tucked right in against the bush edge. The wind would come over the saddle and the smoke would not go up the chimney but just stayed in the hut. It was worst late in the season, at the straggle muster; the condensation on the inside of the hut would thaw and drip on everything.

> We were all out there in the hut late one afternoon in the main muster and all the dogs started to bark. Normally if something came round the dogs would bark but they had a strange bark, and Bruce walked out of the hut and said, 'Good God – we've got a couple of sheilas up here.' Of course everyone laughed. And he said, 'Fair dinkum, there's a crowd of people coming up here.' Well this was miles out – 30-odd miles out, and we thought Bruce could be right, so someone jumped down off the bunk and looked round and here's six or seven of them – jokers and girls – they belonged to a tramping club. We'd all been out there for – well, at least a fortnight anyway, and no one had had a haircut or a shave or a decent wash or anything. We must have stunk like billy-o. Of course everyone's tearing round – 'How's my hair?' how's this, how's that? – and acting the goat. And it finished up they stayed the night. There was one spare bunk in the hut one of the chaps used and they put up two tents.[40]

The Right Hand Fork was 'a bugger of a place to muster' – swampy, and with quicksand in the creek, so they really had to watch that their horses did not become bogged. The boys would go down to get dog tucker, and the valley would be full of deer. 'The Jap deer – the Sikas – were always just on the edge of the bush and they'd give a startled sort of whistle, and go a couple of chain into the bush and then stop. They'd be looking around, watching you, but as soon as you fired a shot, the deer feeding down on the river would come up and there'd be like 60 deer, over a huge length of the river, coming back up into the bush. You could only get a couple, and they're gone.'[41]

On the top of the Boomer, across from the Tikitiki, there is a little patch of bush down towards the Rangitikei, opposite where an old character known as Turkey George used to live on Springvale, on the western side of the river. He had a little house and he worked there, but he used to come over to get his horses from Ngamatea. 'He'd take the pick of the quiet ones, because he was very old.' Once when Ray Birdsall and Ash Watt were mustering the Boomer the fog was right down and they could not see a thing, so they sat down and boiled up. Ash had his little hillbilly with him full of water, ready for just such an occasion, and Ray had no choice this day but to sit and wait for the fog to rise.

> You had to go down a very steep spur to the Rangitikei and muster around on foot and the chap at the top who led the horses, he'd muster the top as well. We'd been sitting there about half an hour and had our cuppa and suddenly the fog lifted from where we were to about one or two hundred yards below us then it was just solid. Suddenly we could hear this creak creak, rattle rattle and up the spur came a line of horsemen with their packhorses, coming to shoot the Boomer. Found out later they were the proprietors of the garages in Taihape coming for a weekend's shooting. They hadn't seen us and suddenly about 40 dogs spotted them and they just bolted down the ridge and after them, and these chaps just wheeled about and took off. I suppose they thought they were wild dogs! After about half an hour the fog cleared down the bottom and we could still see these chaps going like mad up through the other station – Springvale. They'd crossed the Rangitikei and were going for home as fast as they could go. They must have realised it was us up there, surely.[42]

Wild dogs had been a problem in the back country since R. T. Warren's time in the early 1890s and they were still a problem, particularly around the Boyd. That was a popular hunting area with reasonably easy access from Taupo, and lost hunting dogs soon became a menace. Frank Brady and Bruce Atchison were mustering the Harkness when they suddenly heard dogs barking, then saw them running down the track towards them.

> Big Bruce says to me, 'I'll stay behind this tree, Frank. You go back two or three hundred yards and bark at 'em.' They were rushing along towards us and when I retreated they'd come further along, and here's Bruce hiding behind the tree waiting to shoot them when they came along past him. But he didn't have much luck. He fired a few shots and may have wounded a couple, but the cheeky things – when we broke out at the top, here they were chasing a deer. They used to have high jinks tearing after deer. I don't think they'd get

Mustering off Kaimanawa face; Mt Michael above Forks Spur ridge and Haumingi bush; Kaimanawa hut in the little clearing at the edge of the bush.

many of them, but they'd go after our sheep. We were bringing sheep in that were torn to pieces by the damn things. You'd hear them howling across the river all night like wolves, and setting your dogs going.

On another occasion Frank came across a wild dog away up the river beyond the Right Hand Fork.

It was a big pig-dog looking thing, bit of bull mastiff or something in him. I set my dogs alight – I thought I'll sool them on him then I'll get a stick and deal to the damn thing. Found a big manuka stick and the dogs were all packing this thing. When I shot in to kill him, the dogs let go – there was only one dog stayed hanging on to him. Well, he grabbed this good heading dog I had and just absolutely crunched his hind leg. Of course I forgot about the wild dog then – my main worry was my own dog. I tied him onto the saddle and rode back down – I finished mustering by about half past twelve and said to Morrie – 'I'll have to take this dog in'. Tied on the front of the saddle his leg was still flapping, giving him hell, so I strapped him across my chest and I rode all that night to get to the station. It was about three or four days before the Hawke's Bay Show, and I brought him down to Hastings. When I came down again for the show they said they were going to amputate the leg because gangrene had set in.

Going in to muster the Tin Kettle; Frank Brady, Guy Brittan, Dave Wedd.

Luckily a friend suggested taking the dog to Waipukurau where the vet said they could save him: they would operate and put the leg in splints.

> I got him back and he came good as gold, once I got the plaster off. They saved his leg though he wouldn't use it for a long time. But this dog was mad on retrieving sticks from the water and he'd get out and he'd swim for hours. So I used to take him down to the hydro dam at Ngamatea every night and throw sticks in. He'd swim, and he started using his leg and it sort of got him going. Up till then he'd carry it all the time, but once he got in the water he had to use his leg to swim, and away he went. No problem whatsoever after that.[43]

The muster that began early in October had to be finished by Christmas, before shearing. The sheep from the Mangamingi, the Boyd and the Harkness would be brought in to Golden Hills to the wire-netting holding paddock the Tom Harkers – Young Tom and Old Tom – had built; then the Kaimanawa would be mustered into Golden Hills. There was a lot of walking in the Kaimanawa–Tin Kettle area to the Michael, about ten days at a time. They would ride out from the Kaimanawa hut and one man would be detailed to lead the horses back. 'We used to tie them head to tail – there's a real knack in tying horses head to tail, with the reins through the tail – so a horse could really pull back and fight on that tail but you'd always be able to release it

Drafting at Golden Hills yards.

with the special knot you used.'[44] The whole mob would then move down the Taruarau Valley and through a gate in the old rabbit-proof fence into the Swamp – the 18,000-acre holding paddock. The rabbit fence, from the Rangitikei to the Taruarau, which divided the front country from the back, was the only fence beyond those around the domestic paddocks – around the woolshed and homestead. The Manson, Log Cabin and Hogget – the eastern country – would be mustered into the Swamp with the others, then cuts, small mobs, taken in to the station for shearing.

Although there were creeks through the Swamp block, and treacherous bogs, there were also natural walkways across a dense red weed that quaked and shook but supported man and beast. Lawrence would turn the young horses out into the Swamp as two-year-olds and they would learn to pick their way across it so that when they were broken in as about five-year-olds they knew where to put their feet and could be led across the Swamp. 'There was the odd one that slipped, and we'd tie it onto the tail of another horse and pull him out – there's terrific strength in a horse's tail. Ninety per cent of the time you could lead them over, if you picked the spot where the red weed was. You'd see it and jump off and lead them across.' The sheep kept away from the swampy parts – they did not learn the trick of negotiating them – and the musterers would muster around the bogs.[45]

Lawrence sent Arthur McRae to bring some sheep in from the Swamp on his first day at Ngamatea.

Mob in homestead yards; Otupae range in the background.

They were tough blooming sods – not so much hard to work, but totally different. You had to keep off them. First time I'd handled halfbreds or merinos. You had to learn how to move them and learn fast too, otherwise you were in trouble. I had a couple of good dogs, leading dogs used to that sort of thing from the road when we were droving. They knew to stand, and lead off stock, but crikey you could get into trouble. I remember going for this mob and they all ended up in the bog – my first job. The paddock is the size of an average station. I went out there, whipped the dog round. He went off like an east coast huntaway, barking – and I spent the next three hours trying to pull sheep out of the flaming bog. That was lesson number one. I learned quick then.[46]

Once out on the muster there was no way of communicating with the station – no way of letting the cook know the muster was over and a horde of hungry musterers were about to descend on the cookhouse.

We might have been out ten days, longer. So we used to light a fire, put up a lot of smoke – way out towards Golden Hills. We'd be careful where we lit it. It didn't always happen – just if we were early coming in. Smoke signal – they'd see it from the station. Cook would get a bit grumpy if we all turned up there and there was no tucker, so we'd give him a bit of warning. Some of the cooks you couldn't please them though. We had seven cooks in six months one season. Some terrible cooks, and some damn good cooks – but they didn't stay long. You get a lot of drunks out there.[47]

The merino ewes were kept around the station, the English Leicesters down at Pohokura, and there would have been a break in the muster for docking; there were yards down at Jimmy's and at Pohokura. This was shepherding but the men were still classed and paid as musterers, and working seven days a week. The distinction was not between shepherds and musterers, but between 'the boys' – the dog men – and the general hands, who were mostly older and pretty hard case, to say the least. And there were those among them who thought the shepherds were lazy: 'they might be good with dogs, but they won't do anything else'.

Once the main mob – the wethers – were in from the back the boys would muster the hoggets off the Otupae block on the south side of the Taihape–Napier road; they can remember starting at half past two in the morning, and arriving back at the station after ten o'clock at night, and still going out at half past two next morning. The hoggets were shorn first, and they all had to be dagged before anyone could think of going away for Christmas. When there was a stable gang for a few years they would take it in turns to go away.

> Lawrence would say, 'As soon as the hoggets are dagged you can go.' There was some bloody fast dagging went on at times – sometimes we had four or five days to do 3,000 to 4,000 hoggets. Sometimes we'd have a day or two days. They still all got dagged, though. The shearers would start straight after New Year, so we'd have Christmas and we were always back for the New Year. We'd have, say, four days off – but if we never had the hoggets done, we'd be back Boxing Day. And if we had a bad year, a wet season, we never got the mustering done till after Christmas.[48]

Dave Withers was there one Christmas and they went to town in Jack Roberts's vehicle:

> He had a little Humber 80 ute, with a cover on the back, so we took his wagon, went to Taihape to the Gretna, had a few beers, and got home all hours of the morning. The dagging had got held up a bit – it was either Boxing Day or New Year's Day and Lawrence had us at work in the dagging pens, and we've all got these terrible sore heads, hanging over sheep with hand shears – he wouldn't allow the machines. And he'd come and stand over you and make sure it was all done right. His favourite saying was 'Bboy – leave the wwool for the shearers. Just take the dags.'

The old 10-stand Ngamatea woolshed.

Bruce Atchison and Guy Brittan were an unlikely combination – Bruce a rough diamond from a farming background, Guy with a Christchurch private school education – but they worked well together. They were working with Lawrence one night at the end of shearing, penning up his precious merino ewes and lambs.

> All the others were weaned before shearing but not the studs. Mum and lamb had to be together overnight – and we couldn't actually get them in without using the dirt pens down below. But he didn't want to use those on the studs because the dirt might get in the wool.
> We'd had a good shearing that year and everybody used to get a bit tired, and we'd be penning up after tea. So the sooner you get up there and get them in and shut the shed the sooner you can go home and get to bed, because you'd most probably be out at three o'clock the next morning – or earlier. Well, we buggered around and we buggered around in that shed there for a long time. We shifted them and we reshifted them, we did this and we did that. The boss was determined to do it his way – and it wouldn't work. So in the finish I said, 'For Chrissake go home and I'll bloody well do it myself.' He swung round and he glared at me – he had a real glare – he'd line somebody up – and he glared at me, and old Guy Brittan turned around and faced the wall. And the boss looked at me – and he said, 'Good night', and away he went. Guy let out his breath and he said to me, 'I reckon he was deciding whether to bloody sack you or knock you down.'[49]

Loading out wool, 1960; about 26 bales per truck.

But Lawrence knew when the boys were at the end of their rope, and he was not going to sacrifice his head shepherd for the sake of getting his own way. And the boys got the job done – without his help.

Bill Jones did the shearing at Ngamatea for years, starting with the main shear in January 1958. He christened the new woolshed in 1984, and went on for a couple of years after that, becoming an institution on the station. He kept his shearers employed all year round and could send or take a gang for almost any job: docking, fencing, dipping. He would take his cooks when needed, and put down a hāngi for a special occasion. He had a good memory and would remember all the musterers and station hands. Arthur McRae knew him through rugby in Wairoa when they were 'young fellas' and they remained good mates from then on. 'When I was droving we used to go past his place – in packhorse days – and he'd come out in a truck and pick us up. He had a big complex at Fernhill for his gang. It was great – we didn't get many breaks. He'd take us back in the morning to the mob. Have a bath, do your washing – real luxury. No more creeks or horse troughs!'

Everything was shorn, lambs included, and the sheep were dipped 'just about off the shears', through the old wooden swim-through dip. It was 'a beggar of a thing' to work, because the sheep came downhill into the dip, through the dip and into the draining pens.

> In those days the dips were arsenic based – you could smell them 400 or 500 metres away from the dip. The old ewes started sticking their feet in – didn't

Dipping; George Duffel, general hand and Peter Cameron. The old kauri swim-through dip, held in place with wooden pegs, dated from before Watherston's time.

want to go. We were dipping and Alan Bond and Peter Looker were on the crutches, and Alan missed pushing the sheep under and he went in. He battled his way up and was in among all the sheep and thrashing feet and foam and floating shit and the rest of it, and when he got to the surface a voice said, 'Hold your breath.' Luckily he did, because a crutch was placed on the back of his neck and down he went for a second time. He's still alive to tell the story – but I bet he didn't tell you that one![50]

After dipping, the hoggets went down to Pohokura, the ewes were all retained on the front country, and the wethers were returned to the back country and turned out onto all the blocks until the straggle muster. A couple of musterers would take the mob out, travelling light with one packhorse carrying their tucker and bedrolls. It was much easier taking shorn sheep out than bringing woolly ones in; 'they'd motor along, and that's why they only needed two men – three at the outside'. It was still, however, a five- or six-day job; Ray Birdsall described it.

> Lawrence was a bit of a tough man. They'd be shearing, might have 4,000 or 5,000 wethers shorn. About 3 p.m. he'd say, 'Dip them.' They'd dip them, and he'd say, 'Take them out to the wire yards.' You went over the Woolwash,

to the end of Lake block, just over Poverty Creek and you'd get back ten or eleven at night, for tea. Then you went back next day and picked up the mob. Half would go to Golden Hills overnight then to the Boyd, Mangamingi and Harkness; the other half over Peter's, over Kaimanawa face, across Kaimanawa Creek and into the holding paddock at Kaimanawa hut. You'd hold them overnight and next day spread them – some to the Dowden, some on the Zigzag, Tin Kettle, down to Forks Spur, some onto Mangamaire, some up the Michael – big climb, the rest on the lower Michael. That fellow would walk about 20 miles and climb up out of the Mangamaire onto Forks Spur and find his horse – tied there for him by the chap who took sheep to Forks Spur.

When we mustered the Michael two fellows would take sleeping bags and a bit of tucker, walk down Forks Spur and camp at the foot of the lower Michael, by the Rangitikei River where the Mangamaire meets it. I woke up there one time – the last time I did it – with Peter Cameron, covered with snow. And this is the main muster. Howling blizzard, mustered all day in shorts, got to Kaimanawa hut – and no one else turned up. They were supposed to come with our horses, but they'd all thrown it in – the weather was too bad![51]

Lawrence would decide how many sheep would go back onto each part of each block and as the musterers moved forward they would be counting off a certain number into a certain area. If the sheep were just pushed out onto the block and left to spread on their own, they simply would not drift very far. Merinos and halfbreds are gregarious and they would pretty well stay put and eat the area out; so a few would be put over a creek, a few onto that face, 20 or 30 would be left in that gully, a dozen up that ridge, and so on until the mob was spread in little pockets all over the block. They stayed more or less in those pockets until the next muster when the musterers would have an idea of how many to expect off each area. As more sheep were shorn they might be taken back to Log Cabin or the Hogget, or through the Rock Camp area and on to the Manson, where about 900 six-year-old wethers would be turned out.

By the 1960s contention was growing about the grazing of the back country, especially that around the upper Ngaruroro. In the years leading up to the establishment of the Kaimanawa Forest Park in 1969 the Crown talked about the need for fencing to exclude Ngamatea sheep, but Lawrence was not enthusiastic about such ideas. He would ask the boys what the feed was like out at the Boyd. When they told him the tussock was taking over – in the holding paddock it was 'as high as your saddle' – he said, 'You'd better do a Bryant and May trick, but don't get caught.' So as they were coming out with the pack team and the last of the packhorses passed through the holding paddock gate, one of the boys shot back and dropped a couple of matches. But one of the packhorses broke back and they had a job getting him out of it and away before the spotter planes came over to see what was going on.

The end to the grazing of the Harkness came in the early 1960s as a result of fire, and the Harkness became part of Kaweka Forest Park. The musterers reckoned the best stock came off that country, which used to run 1,200 two-tooth ewes. Lawrence never put the two-tooths to the ram and there were no wild rams out there, but despite that a few young ewes always managed to get in lamb, coming or going. The Harkness was a lovely valley, good warm country, but a terrible place to muster off because of the huge golden tussock, the biggest on the place: you could not see through it and it was so high you could tie it across the front of your saddle. The worst area was right on the Oamaru–Ngaruroro watershed, 'a small area – but just when you got the stock there it beat you every time'. At one muster-off they had had enough of it and someone dropped a match. 'It wasn't much of a fire – wouldn't be 3 acres, just a bit of a good clean out' according to one observer. The surrounding bush was green and it was only the saddle that burned, but in about 20 minutes the Forest Service fellows were there and that was it – no more grazing the Harkness.

> They sent Dad the bill for all the aerial inspections and there was a bit of argument about it, so he decided he'd pay it, but he didn't pay the exchange on the cheque – just a bob or two – and they kept sending him a bill for that and it kept on and on and on, and they must have spent goodness knows how much in stamps sending him this bill. They finally gave it away – it was like trying to get blood from a stone. I think it was the head of the Forestry at Kuripapango, and he met my father on the road one day and said plaintively 'Why do you light these fires, Mr Roberts?' and Dad said, 'To keep buggugers like you in a job.'[52]

The straggle muster of the back country varied a bit over the years. In Bill Cummings's time most of the wethers were held round the place for a while and the back country would be straggle mustered before the main mob was turned out. 'We'd get quite a few – hundreds, not thousands. Sometimes there wouldn't be that. All depends on how the main muster went, and how many you got on the Manson where the wild sheep had led them off. We'd muster the stragglers from out around Golden Hills into the holding paddock there and drive them to the set of yards at the Swamp, then bring them into the station from there.' In Ray Birdsall's time the straggle muster took place in April–May and the whole flock came back

Winter 1959; halfbred ewes heading down to Timahanga and Pohokura; Ngamatea homestead in centre.

in. It was often the end of May or the first week in June before the Manson was straggled.

> There'd be delay after delay, rivers high, horses being swept away. The boys wanted it done at the end of March, but there were always other jobs. So there were two full musters – the stragglers joined up with the rest of the mob. In the first muster the sheep were very weak and slow, from winter. But at the second muster they were all better fed from grass growth and you could do it all twice as fast. They'd shear perhaps 1000 stragglers and wig others – take the topknots off – to stop wool blindness. Young merinos, halfbreds, grow a lot of face wool. They only wigged the young ones.[53]

A bit later they built a set of yards at Golden Hills so that when they straggle mustered they did not have to bring all the sheep in. They would muster the back country into the holding paddock and through the yards, the shorn sheep would be taken back and spread out again, and the woollies brought in to the station. Everything would probably still have to come in from the Manson to the east, and Peter's to the west. Then a small shearing gang would come up to Ngamatea to do the straggle shear – anything from two to six shearers and 1,000 to 1,700 sheep in the years between May 1952 and May 1963 according to the records.

The crutching muster, which came in June–July, after the main mustering season had ended, involved only the ewes and other sheep on the front country and down at Pohokura. One or two of the musterers might stay on or come back for it. Alex Lindsay came back from Mangatapiri in July 1956. 'We couldn't do anything. It was the worst snow they'd had in

20 years – it was over the fences in some places.' Craig Hammond, Don Hammond's son, arrived on a bright mild Sunday in July 1977. It was his first day in a new job, one he had looked forward to for years. 'I'd gone up with Mum and Dad and we went for a drive around with short-sleeved shirts on. Monday morning there was six inches of snow. I wondered what the hell I'd gone into. I thought I was in Canada! They were crutching – the gang was there, shed full of sheep, but they had to let them out.'[54]

Some young fellows got their start by coming for the crutching muster and staying through the winter. Dave Withers and Dave Wedd arrived up there a few days apart in June 1959. They had met through dog trialling around Hawke's Bay and got on well, but neither knew the other had applied for a job at Ngamatea. Although it was cold and bleak, with plenty of snow, Dave Withers loved that first winter.

> After the crutching muster and the hoggets had all come in and been crutched Mr Roberts wanted Dave and me to take a mob of a good couple of thousand out through the Road block onto the main road and deliver them down to Timahanga so they could be spread out down in the warmer country for the winter months. Big mob on the road – but we were pretty confident young guys, we had these good leading dogs as we thought. So we get out on the main road and it's snowing and as you ride along the snow builds up in the crook of your arm. Coming over the top of the Taruarau hill to dip down to the Taruarau River it's round about lunchtime and there's this little clearing. We had our lunch with us and we thought we might just push in there out of the rough weather a bit, might get a little fire going. We'll have lunch there . . . we'll just leave these good leading dogs in the lead and they will hold the mob up . . . we won't be long. So we nestled in and got quite cosy, ate our lunch, relaxed, got warm, it's still snowing quietly and we know we've got good command of the mob with our good leading dogs, nothing to worry about. And we quietly dozed off, didn't we. And we wake up. Where are we? The mob? Oh, they'll be all right. We look a bit further, behind our huntaways, and there's our leading dogs, snuck back through the mob and they're lying down with us all nice and cosy. Well. All we could do was go forward, pick up what we could of the mob, and not knowing the big gorgy country in to the right of the Taruarau, leading up onto the Otupae Range area – that's where a lot of the hoggets had gone – we just took what we could down the road and delivered them to Timahanga. The rest were all over the place. We were very embarrassed young shepherds going back to Mr Roberts and explaining the situation.

Lawrence took it well. Anything lost up there was lost on Ngamatea and would be found next season; and when that came around, the two young shepherds and their dogs were ready for the challenge of the main muster.

Chapter Five

Incidents and Accidents

We're full of stories and they need writing down.

One well-known Ngamatea story concerned Jim Crowe, a fellow of dubious provenance who had turned up at the station early in 1941 and been sent out to the Tikitiki bush to cut firewood. Ngamatea went through a lot of firewood – they cut up to 200 cords a year. Jim Crowe was not much of a hand at the job but he was working with Tom Wilmot, a first-class bushman, and one evening about eight o'clock Tom turned up at the door of the homestead, having walked the eight or so miles in to the station, and told Gordon Mattson that 'old Jim had kicked the bucket' – he had simply sat down under a tree and died. Gordon rounded up the tractor driver, Dave Clifford, who was the house-cook's husband, and the three of them took the truck out to the Tikitik. They were able to drive up part way into the bush then walk to where they found old Jim.

> He was sitting against a stump with a hat pulled over his head – you'd think he was just having a sleep. But he was dead all right. The tractor driver was a bit dickey – he kept in the background, but I said, 'Come on Dave, you grab his legs.' I got him under the armpits and he stayed in the sitting position and we started to carry him down the steep bank. His arms were crooked and before I could say to stop his arm caught in a bush lawyer and he whipped back and hit Dave and Dave gave a yell and nearly dropped him. We got down to the truck and put him on the back and there was a tarpaulin rolled up at each end, so I put one roll on his knees and one on his chest hoping that would straighten him out before we got home and we stacked some firewood against him so he wouldn't roll. When we got in to the station, sometime after midnight, I drove the truck into the stables and closed the doors on it. We rang the police and they asked us to bring the body in – they wouldn't come out.[1]

Next morning Winnie, up early to stoke the fire, saw the stable doors open and the truck backing out so she called Gordon to get down there and see what was going on.

One of the station hands was driving. His name was Reg Gillan but the boys called him the Cactus Kid because he was a townie who thought he was 'away in Texas'. He'd left his pony in the shed all night, which he shouldn't have done, but he wanted to get it out early and he'd seen the truck and promptly backed it out. Gordon said, 'You shouldn't have done that – there's a dead man on board there' and Cactus of course thought he was pulling his leg and said, 'What do you mean, a dead man on there?' 'Come and have a look,' said Gordon – and he lifted up the corner of the tarpaulin and Cactus looked. 'Jim Crowe,' he said. 'How did it happen, Gordon?' and Gordon told him how it had happened. 'I can't believe it,' he said, 'Jim Crowe . . . working yesterday . . . and dead today. Oh Gordon. Poor old Jim Crowe, poor old Jim Crowe – lying there . . . just *lying* there. He always was a lying bastard, Gordon.'[2]

Gordon and Tom drove the truck into Taihape with old Jim on the back and the police told them to take him straight up to the hospital morgue. The staff there wanted to know a few particulars about his family, his background, but with a name like he had been using neither Gordon nor Tom could help. The police could trace no relatives so Gordon found himself arranging a funeral for someone he knew nothing about. 'Make it Anglican,' he said to the funeral director, 'and too bad if he's Catholic.' He rounded up Colin Kirkpatrick, the gardener, Dave Clifford and Tom Wilmot to go into Taihape with him in his old Model A and bury Jim. But first he had to go out to the Tikitik to tally up how much firewood Jim had cut and calculate what he was owed for that, then deduct his store bill – and take the rest in change to pay for the funeral. 'I had it in my pocket and when the minister finished the service I handed it over. It was about 35 bob.'

Some of the memorable Ngamatea stories were passed down through a succession of musterers, and Ivan Chapman knew a lot of them.

It was the year after I left. My mate Ken Fraser had the horses tied together. All the musterers had gone out and got off their horses and went on their walking beat, and Ken was to take the horses back. I know the exact spot where they went down, because it was a terrible track round the side of a hill and a fair sort of a drop. One of the horses pulled back. He had them tied head to tail, and they went down over the track and Ken went down with them. He was lucky to come out of it alive. He wasn't injured, and his horse never got hurt, but they had to shoot one, maybe two of the horses. Well, he lost a lot of skin – and he smashed his good saddle – an Association rodeo saddle.

But he did break a leg in the stockyard – that was after my time too. They put a splint on it and someone took the back seat out of his car and they took

Firewood coming in from the Tikitiki bush.

him in to hospital in Taihape. Margaret came down from the house with a bottle of whisky and he'd had a couple of whiskies before they put a mattress in the back of the car, and Margaret was going to take the whisky back but Ken said, 'Oh no, don't take that – that's the only thing that will keep me alive.' By the time he got to Taihape he'd had quite a few shots of this, and the old matron at the hospital give them a fair roasting because she couldn't do anything while he's got alcohol in his system, so they had to dry him out first. They reckon it was one of the worst breaks they'd seen in Taihape hospital. Ken went to stop a beast at the gateway apparently and got his foot caught behind the strainer post and it twisted it round – it wasn't just a break, it was a twist, and it shattered the whole leg.

I went up to Ngamatea just afterwards and got three or four of his dogs and brought them down and looked after them while he was in hospital. And he ended up marrying the nurse on his ward. She must have been a good nurse I reckon.

Ray Birdsall always spent the break between mustering seasons doing a lambing beat down in Hawke's Bay. One Saturday night in 1955 he turned up at an Ingleside at Omakere, and his father introduced him to Maxie Mackenzie, who had been his nurse in Waipukurau hospital and had become a great favourite with the Birdsall family. Ray and Maxie's fate was sealed from that moment. They had the supper dance then sat in the supper room for an hour, talking – about Ngamatea. According to Maxie that first conversation was 'absolutely marvellous – any talk of Ngamatea and Ray's eyes would light up. He adored the place.'³

A few weeks later Ray went back for the mustering season, and Maxie wrote to him as promised – and remembered that he had asked her not to number her letters. 'Apparently one girl had done that and the boys would change the numbers around when they picked up the mail.' Maxie wrote voluminous letters, and Ray spent his time trying to collect them before anyone else saw the incoming mail. As head musterer he could arrange it so that he was the one who came in with the packman to get fresh hacks, stores, bread and mail. The night before they came in to the station he would get a candle and sit in a corner in a dark smoky hut with a little pad on his knee and write a meagre page or two to Maxie. All the boys knew what was going on of course, and they gave him beans. If they ever saw him reading one of his precious letters they would all find a reason to walk past him and one would say, 'Come on Ray, read us the sports page.' But far from sharing the least confidence with anyone, Ray did not even keep the letters – he burned them, though he did not tell Maxie that.

It was soon arranged that Ray would come down and have Christmas dinner with the Mackenzie family. In Christmas week Maxie went to the post office with her father to clear their mail-box.

> There was this lovely big Christmas card – and in those days a big Christmas card with a padded Father Christmas was 2/6, and a boy could go and get a meal with fish and eggs and chips, a really good meal for 2/6. I was surprised, and I opened up this big envelope there and then, at the post office, and the card said 'To Maxie – In appreciation for all those nice long letters, from The Boys at Ngamatea'. I just dropped it and wailed, 'That's it. That's it. It's all over. It's finished. Letting the boys read my letters. . . .' And Daddy said, 'Oh now, now, now, Ray wouldn't do that.' 'Oh yes,' I said, 'he's done it all right.' I was *so* upset.
>
> When he arrived for Christmas I opened the door. He said, 'Hello', very eagerly and I said, 'Hello', very coldly. 'What's wrong?' 'You know what's wrong – letting those boys read my letters.' 'What are you talking about?' 'You just wait there' – and I shut the door and left him standing outside while I went and got the card. He looked at it, read it and said, 'That bloody Mott!'

Despite the boys at Ngamatea and the tyranny of distance, Ray and Maxie managed to continue their courtship. They saw each other seven times in the seven months that they were engaged. It was long before the Annie was sealed and it would take Ray over three hours to get to Waipukurau. 'He'd get there about six o'clock and go back about ten the same night. Usually we just sat in the car outside the nurses' home. They were wonderful days!' They were both delighted when the Roberts family accepted an invitation to their wedding in August 1957.

Time off was hard to come by in the mustering season. Some years later Paul Thomas and Dave Withers asked the boss for a Saturday off so they could go back to Wairoa – Paul for a friend's wedding just out of Wairoa and Dave for a twenty-first birthday in town. They left Ngamatea on Friday night, after work, after dark, in Paul's cantankerous truck, got to Napier and could not find any petrol. In those days even the gas stations closed at 6 p.m. and neither the Automobile Association nor anyone else could help them, so they were stuck in Napier for the night. They made it to Wairoa next morning and Paul went on to a very good wedding.

> Part of the agreement was that we actually had to be back at the station on Sunday. We got back about ten in the morning and old Roberts looked at me: 'You're l-l-late'. We were supposed to have been earlier than that. We were marking calves in the yards just behind the stables – and tossing those bloody things. They were yearlings, they weren't calves at all. Withers and I were feeling very hung over from the night before, and we were sent in to bat with these calves. They *all* had to all be thrown – it wasn't a case of putting them up the race or anything like that. And the big secret with marking calves was you had two guys, one on the head and one on the flank, and you *must* work as a team because if the guy grabs the head at a different time from when the other guy grabs the flank and your timing is not quite right, you're going to miss out, and the calf's going to get away.
>
> Look, Dave and I were kicked from pillar to post and I think the old boy had a grin from ear to ear: 'Serve the buggers right'! Some really were yearlings and we had to actually tie them to a post. There's a post in the centre of the yard and you get a rope around them and wind them in to the post and cut them and brand them there, standing up. Well, when Withers and I went to bed on that Sunday night we were bruised from head to toe.

Some of the really bad accidents on the station involved outsiders. One of the worst was a death on the Donnelly in the winter of 1959. A group of students from Auckland planned to come in from the Napier–Taupo road and tramp out through Ngamatea and had written for permission to use the musterers' huts. The first indication of trouble came late one very frosty night. Winnie was reading in bed when she heard the swish-swish of an oilskin as someone walked up to the house, then a knock on the front door.

> I got up and went to the door and there were two young fellows, just about exhausted, and they just sat down on the verandah, where they were. I

remember afterwards the frost was melted where they were sitting. They said one of their party had been killed out on Makorako, the highest point on Ngamatea. I brought them inside and gave them a meal and they rang the police station and the police immediately rang the head of the newly formed Search and Rescue Association. His reaction was 'Oh yeah, a student fallen over a hill at Ngamatea?' And the police said, 'Yes, get out your association and get cracking' and the head of the newly formed association said, 'Not at this bloody hour of the night, thank you.'

However, the next morning they were arriving at Ngamatea and before long there were quite 40 Search and Rescue people come to help. These students had to go all the way back to Makorako to show them. These two boys were the advance guard and the others were still wending their way in. We don't know why they had decided to climb Makorako on their journey through. We'd had terrifically cold nights and there was a bit of snow on top, frozen into ice, and one poor lad had slipped and fallen about 100 feet down and was killed. We had one horse that was very quiet – he had been used to bring hurt people in at times – broken-leg men, and old Bluey was brought out again this time to bring the body in.[4]

Dave Withers and Dave Wedd were on the station that winter, and Margaret was home from Massey. All three rode out towards Golden Hills and took quite a few horses to meet up with the Search and Rescue party, most of whom had gone on foot. One of the students and one or two of the police could ride, and they had gone out with the packhorse to bring the body back to Mangamingi, and from there to the Boyd airstrip so it could be flown out.

It was the last crossing of the Taruarau before the Golden Hills we met them – about ten of them in that group with the police, and half of them had never been near a horse. . . . They had these big boots with tricounis – you know, those metal studs for climbing boots – and they got on these horses and put their feet in the stirrups and the boots were never going to come out of those stirrups, and several of the horses just took off and they never had a clue how to stop them. Luckily there was quite a hill there back up towards the Lake block and they pulled up. But it was pretty hair-raising for a while. Some walked and some rode. The students were out of it by then – it had all taken a few days.[5]

In May 1967 a Cessna 172 with a pilot and two young shooters on board crashed at 4,000 feet on the top of the Hogget. They had been heading for the Boyd the evening before, but it was getting dark, the fog came down and they could not see the airstrip, so the pilot carried on until he saw the flat top of the Hogget and made a good landing among the mingimingi.

> They put a little pup tent up and at daylight the next morning there was a stag standing looking at them so they grabbed a gun and shot it and carved all the meat up and put it in the plane too, and that made it too heavy. They cleared a lot of the mingi and rubbish and they got off and got airborne, but there was a bit of a hill in front of him and he couldn't quite make it so he turned and started to come down this big gully, heading towards Golden Hills. But he must have got in a down-draught or something and he went nose first into the end of this hill and banged into this big rock face.[6]

Ray Birdsall had taken over as manager of Ngamatea a few months before and late that morning was welding a plough in the blacksmith's shop when 'a face appeared in the doorway – all swollen and absolutely black. You could hardly see the eyes.' This young fellow was not burnt but his face had taken a pounding when they crashed. He had walked in about 14 miles to get help. His mate was back there 'all smashed up and his bones sticking out'. He had got him into his sleeping bag, made him as comfortable as he could and left him lying under the wing of the plane. The pilot 'wasn't too bad, just concussed': he had sat him on a hill above the plane with an orange tent as a crash marker. This young shooter had no idea where he was – he had seen a road Ray had had bulldozed about six miles across the Swamp block and he followed that south, thinking he was heading north, for Poronui. Ray phoned the police and said they needed a helicopter out there, even a plane to find the crash site for a start. The police were less than helpful: they could possibly supply a plane, then Ray could send horses out to bring the injured in. Luckily this was going over the air, and a plane in the vicinity heard it and came in to see if he could help. Before Ray hopped in the plane he sent three of his shepherds off at a gallop towards Golden Hills with blankets and torches. He knew the wreck would either be on Peter's, to the west, or the Hogget, to the east; they would circle it when they found it so the boys would know which direction to head.

> I saw them from the plane, galloping out – and they were more than half way there by the time we found the crash and circled. They tore up onto the Hogget – they were musterers – they knew the country. We came back and just as we landed another plane came in – a top-dressing plane, the one the police had sent. He couldn't land anywhere near the crash so he went back to his top dressing – and it took till nine that night before I could coax the police to give us a helicopter. It was terrible – absolutely terrible. I said, 'This

joker's got his hip sticking out the side of him, I can't bring him in on a horse.' Eventually they said, 'All right' and the helicopter came out. By this time Jack had come up from Timahanga and we put all the vehicles we had in a ring in the aeroplane paddock with the lights on. They had Dr White from Taihape – a good doctor – he'd been in Vietnam – and a policeman and the pilot. Nine at night, pitch black, in May, cold as anything. Away we went. The pilot asked if I could guide him with his searchlight. 'Yeah, I know the country – I'll take you there without a searchlight.' I was counting the ridges on the Hogget and struck the Otutu bush, then there's the torchlight below us, with the boys. We came down on the torchlight. They said, 'We've been here for hours – we can't find the aeroplane!'

They had found the concussed pilot, but he was no longer near the plane. When he saw the boys coming up with the horses 'he picked up his tent and tore away up the Hogget to meet them – and then they didn't know where the aeroplane was – they couldn't find it. And the fellow had been alone all day, with his bones sticking out.' The pilot was sure his plane was 'up there'; Ray was sure it was 'down there'. The doctor and the policeman thought the pilot should know, but luckily the helicopter pilot put his money on Ray and they took off again, heading downhill with the searchlight probing through the scrub. 'I said, "About here", but you couldn't see it, so I got him to land and we went over the edge and down a bit and yelled out to him to tap on something, and next thing we heard tapping and we went down with the stretcher. They put him in a big air bag.' The three shepherds, Jack McLellan, John Bulled and Barry Dunn, had tied their horses up and followed the helicopter down.

> We were there ages – all sorts of injections and things – then took turns at carrying the stretcher up the hill. It was some struggle. He was still lucid. We put him in the chopper – plenty of room – then they said to me, 'Oh we can't take you back – you'll have to walk.' And they left me there – and they wouldn't come back to get me. They said they hadn't the right to do that – 'You're not injured'.[7]

The four of them went back to where the horses had been tied but they had gone, spooked when the chopper took off with its searchlight blazing. Ray and the boys followed, with their 'little weeny torches', down to the first crossing of the Taruarau. 'We had to cross it twice more, and next thing a pack of wild dogs start having us on, right behind us, howling and barking at us, and the more we walked the more they just kept level with us. They followed us all the way through the Swamp, howling and barking all the time. They were close.'

Maxie had her part in the drama. 'Russ, the boy that came in with the black face' was sent over to the homestead for her to patch up.

> I mothered him and got him into a bath and gave him one of my brand-new towels and there was blood all over it when he finished. Poor boy, cuts all over him, not serious ones. He said he was all right, but he was concussed. He wouldn't go to bed so I got two armchairs and put them together and the children rushed and got rugs, and covered him up and I gave him hot soup, and he told me what had happened. I had to run way over to the airstrip and back at one part of it and I told him he mustn't move, and I told the children to see he didn't move. I thought he'd fall all over the place if he got up. I was worried about my patient, and leaving the boys on their own while I was running around in the dark. It was quite a drama. Ray got home so late at night, and I cooked them all a meal. I didn't know where they were or why they were so long.

Ray had had his own drama with a horse when he was a musterer at Ngamatea, and he was sent up to the homestead for Winnie to patch him up.

> A beautiful horse – tall, about sixteen or seventeen hands. Walk! I've never seen anything walk like it. I rode that horse for years, but before I got it settled down it dished me, right up by the Tikitik, along the top there by the Tit, and as it went over the top of me it kicked me in the face, ripped all the skin off, broke my teeth. Anyway I managed to catch it because the saddle had come right over its head and pinned its head and ears down between its front legs. The crupper had broken and that's how it got rid of me. But I rode on – we were mustering right around the back of the Swamp. I had a choke strap from then on for the rest of the day round its throat, back along and onto my saddle. And every time it went to buck hard and get its head down far enough, it would cut off its wind and it had to get its head up again, so it would stop bucking. I mustered all day right round and I got within about half a mile of the hydro dam on the eastern side of the Swamp, and I thought, Oh, you poor horse. It was wheezing a bit, so I undid the choke strap and got on – and it dished me a beaut – and bolted for home. When I got home one of the boys must have said to Mrs Roberts, 'Ray's a bit hurt', so she got me and patched me all up. She was a real trimmer – anybody hurt at all she'd look after them. I honestly think she was the real stabilising force on the place.

An early accident on the station involved Dave Lumsden, a Kuripapango neighbour who knew the country well. He had been shooting on the

Manson and was within a chain or so of the Log Cabin when he fell in the pack-track and broke his leg. Ray White was mustering there at the time.

> Jack Steele, our Canadian packman, found Dave in the Log Cabin in a bad way. He was 6 foot 6 and about 18 stone, but we had a very quiet, very good packhorse and we deepened one of the pack-tracks outside the Log Cabin and next day we backed the mare into that trench so she was a lot lower, and Keith Brooks, one of the musterers, 6 foot 4 and strong as a bull, and two other men lifted Dave onto the pack-saddle – we'd put a mattress on top. We led it down to Timahanga and the ambulance was waiting, backed into a loading bank so we loaded him on. Poor old Dave must have been out there for over a week.

The notorious Napier–Taihape road took a toll on the nerves, and on the station's vehicles, but most of them managed to stay on the road. When Ray Birdsall was manager on Ngamatea, he came to grief quite late one night coming up the Annie. He had been down to the Bay and was driving the station's big four-wheel-drive Bedford truck, bringing back half a ton of potatoes and a load of drums of fuel: this was the late 1960s, in the days before a tanker came through to supply the station.

> The grader had just been through and he must have put all the soft stuff just on the edge. I thought the road was solid and I just went out too far to get around that corner – you couldn't get round in one run – you had to back – and those four-wheel drives don't have much lock. The truck started to slide – over the side of the Annie. Luckily it bellied underneath. It didn't go right over – it was just hanging there – but my load tipped. Everything was tied on, and a big rope round it all. I had to walk back down to Kuripapango, to Rosie's, and a chap offered to drive me back to Ngamatea. Next morning the grader driver brought his grader along and Jack and one of his men, Joe Campbell, brought the 5-ton Austin up from Timahanga and we strapped it onto the side of the Bedford so it wouldn't go the rest of the way over when the grader started to pull it. She just climbed back onto the road, no trouble. I didn't lose any drums of fuel, but I lost a few bags of potatoes – those spuds are still down the side of the Annie. And it wasn't long after that the Army decided to come through with all their gear and they blew that corner up.

There was very nearly a terrible accident when Lawrence was driving the boys down the old one-in-four grade Spiral between Timahanga and Jimmy's. He had four musterers with him – Jack, Peter Looker, Murray Taylor and Kevin Sayers – plus three dogs each, all crammed into the short wheel-base Land Rover, and at the top of the Spiral they were expecting him to stop and put it into low four-wheel drive. But he was thinking of other things and they were all very hesitant about telling the boss how to drive, so they held their tongues. As they started down Lawrence started to stand on the brakes . . . harder . . . and harder. But nothing happened: the brakes were permanently wet from fording creeks – they had crossed two just before the Spiral and they were picking up speed. They reached the elbow at the bottom, and they will never know how they got around it, teetering on two wheels, and all thinking: 'This is it'. But they made it, onto the last stretch of road – and in front of them was a closed gate. There was no way to stop. They took the gate with them, across the bonnet, down into the creek and straight through the boulders. When they came up the other side they finally stopped. The boys sat motionless, stunned, amazed they were still alive. The hot brakes were steaming and the ends of the gate were still flapping across the bonnet. Lawrence turned to the gate-opener and said, 'That's one you won't have to open.' No one even dared laugh.[8]

It was surprising how few accidents there were in the normal course of work at Ngamatea, given the isolation, the rough country, the river crossings, the amount of horse work 'and the silly things we used to do – and we didn't even carry a real first-aid kit – a few plasters, and a bit of that horse ointment, you know, the Rawleighs . . . and a couple of aspirins – that's about all'.[9] But there was always 'the odd emergency, the odd injury. The district nurse used to come out every now and then and give you a jab. Mrs Roberts was nurse-cum-everything.'[10]

Dave Withers managed to break a wrist – 'and that was fooling'. They would play cowboys: stand their horse under a bank and run out and jump into the saddle.

> We were mustering from the Tikitik around the Boomer and we were down in White's block. I'd been pushing some sheep around and ran down the bank, and my horse was standing in the creek and I jumped, but he moved away and I missed the saddle and tipped over into the creek. I thought I'd just sprained my wrist. It was about six weeks later they decided it was broken. I couldn't stand the pain at night and it started to swell up and was quite knobbly. So I went to Taihape for an X-ray and hey, it's broken. I decided to come home to Wairoa and go to Gisborne to have it operated. They took the bone out – it was decayed – and he put a bit of plastic in there. It's more or less good as gold, but it doesn't bend too well.

An early tractor driver, Happy Robinson, who stayed at Ngamatea for many years, was an argumentative character. He and Mac McMahon, the rabbiter and one-time boxer, would start to argue in the cookhouse and end up having a row. 'They were terrible,' said Ray Birdsall, 'they'd be so heated about some little thing, and one time old Mac walked outside first and stood right by the door and when Happy got up and walked out Mac belted him and knocked him flat. He was waiting for him to poke his head around the door and he just hit him.' Some years before that, when John Roberts was there, poor old Happy had turned a crawler tractor over on himself when he was out ploughing the Long Oat paddock.

> We were at the cookhouse one night having tea and Happy didn't turn up so we decided we'd better go and see what the story was. We jumped into the truck and we found him – he was pinned underneath the tractor seat, by the legs, upside down. He had the diesel tank in front of him and he had let the diesel out and it burned his legs. We were frightened to move him because the tractor was going to go over, so we had to go back to the station and get long lengths of wood and stabilise the tractor. Eventually we got him out and old Happy wasn't too happy so we decided to take him in to the hospital. I had an old Chrysler 75, a big tourer. We got Happy into the back seat and every bump we went over he moaned and groaned and carried on. We didn't take too much notice, and about halfway in old Happy decided he wanted to go to the toilet. We couldn't move him. As soon as we started to shift him he moaned and groaned some more, so there was only one thing to do. I bent down and took his boot off and said, 'Here, use this.' We got him settled down again and continued on. I don't think he had anything broken, just a bit of diesel burn. He was very very lucky. That's how easily accidents happen, and you just had to deal with it the best you could. Anyway he got a new tractor out of it – they replaced that one with a new TD9, much wider in the tracks, and more stable.[11]

Mac the rabbiter got his comeuppance when he decided to try his hand at carpentry. Joe Larrington was the station carpenter at the time, and he warned his would-be assistant to watch himself with the saw.

> If you're not used to handsaws they can jump. This was a Saturday, and the saw jumped and cut him badly between thumb and finger, enough so the tendon went. There was nobody home either, so I made a bit of a splint out of some wood and put it on and bound it up and we were taking him in to Hastings hospital, but he wanted to get off at the Fernhill pub. A little thing like a thumb waggling didn't concern him very much. He demanded to get off, so we let him off. When he arrived at the hospital of course he was three sheets in the wind. But they fixed him up and he came back eventually.[12]

Then there was the time, way back in the 1940s when the cowboy was cleaning up some trees in the old orchard, cutting some branches off, and

George Everett with his mates.

he cut himself rather badly. After they'd taken him in to hospital by car Gordon Mattson saw this axe stuck up in the tree, and went to retrieve it. 'When he was climbing up the tree the axe came loose, and he met it coming down as he was going up, so he got clobbered too. Because the car was away taking the cowboy to hospital they couldn't take Gordon in for a while till the car came back to collect the next one.'[13]

George Everett had an accident when he was camped out at the Tikitiki hut in the early 1960s. No one knew for a while because he was a very solitary and self-sufficient man who turned up only when he needed to. He did not even pick up his wages for years: he got one cheque in September 1955 and by April 1958, when he had £1,200 owing to him, Lawrence said, 'Better come and pick up your wages, George, before you own the bloody place.' George was strong and fit, a good shot with a rifle and had 'hands on him like dinner plates'. He trapped and poisoned rabbits and dug up acres and acres of rabbit burrows over the length and breadth of the station. The boys would ride past his camp at daybreak and the embers of his fire would just be going out. He looked after his dogs devotedly. 'They'd get onto pigs and get all ripped up and he'd lay them out on the mat and sew them up himself. He loved them – they were his real mates.' You would see George coming, with all these dogs fanning out around him – 'ruthless killing machines' as far as rabbits were concerned – and cats. There were always a few cats around the cookhouse 'and one day George turned up out of the blue with all these dogs, and all of a sudden there was a yelp or two and cat fur flying around and . . . Nothing – completely disappeared. One less cat.'

He would walk in to the station every now and then for his stores. But one day he failed to turn up when he was expected and the boss told Alan Bond, the station bookkeeper and general factotum, that he had better get out to the Tikitik and look for him.

> I got out there and was calling, 'George, George.' I wasn't going to get off my horse and face all those dogs – lurchers and whippets and bloody things – but eventually he opened the door and I thought God almighty, he looks like a ghost. He'd been walking home at night through the bush and he tripped and it was either his .22, or he hit a log and smashed all the side of his face. It was a helluva mess. He'd just stayed there – I don't know how many days. He was just going to die there I think. I didn't have any first-aid gear or anything so I took off at a great rate of knots back to the station and the boss sent two or three of us off with the tractor and trailer. My biggest fear was what the hell are we going to do with his dogs? By the time we got back he had everything under control, dogs tied up, but he was very weak and he looked a mess. We got the trailer as near as we could and got him onto it and back to the station and straight into town and in to the hospital. They operated on him – he was a pretty sick boy. The nurses couldn't get over him – they had no pyjamas big enough for him and his feet were hanging out the end of the bed. I don't know how many days he was in hospital – a couple – and the hospital rang. He'd disappeared. He'd got himself dressed and taken off, and a day and a half later he arrived home. He'd walked, all the way from Taihape, head all bandaged up.

In extreme weather when the musterers were confined to one hut for days at a time their teams of dogs would get as bored and frustrated as their masters. The gang was snowed in on the Manson one season for ten or twelve days. They only went out of the hut for more firewood and water and the head shepherd would tell each one when he could go out and let his dogs off.

> You couldn't go and let them all off together. Just imagine them – they're full of energy, they're jumping out of their skins. Living on that wild venison meat they were like wild dogs, and they'd fight just for something to do. They go nuts when they're not working. We could let off say two teams

together, so you could control them. You'd each get maybe half an hour out there with them. They lived in holes dug out of the side of the hill with a stake to tie them to – that was their kennel. We all carried those little light dog chains.

At one stage, thinking the day was coming right, a lot of us let our dogs off together, and that's when we learned. There was a dog scrap, and once they start on one, they're all in. I remember the boys wading in with anything they could pick up – pack-straps, the axe, or the shovel from the fireplace, whatever – just to back them off and split it up. They were ferocious. A pack like that, even if you had only 20 dogs onto one, it's a power to reckon with. Normally, when they're in work, they're not like this. You'd take your beat and when we'd meet up, several of us, a mob of dogs, no problem. And when you were riding together, no problem. It was just really the fact that they'd been cold and tied up.

That time at the Manson, we'd cut a reasonable-sized red beech, and as long as it would fit inside the hut and we could close the door, we'd poke the end in the fire, and as she burned we'd just keep nudging her into the fire. We used the log as a seat and a table while it burned through over the period of the day. Tempers got a bit frayed that time – but we learned from it – when to speak, when to bite your tongue, and how to get on in a small confined area with a group of men.[14]

In the 1950s a fisherman drowned in the Rangitikei River, on the Ngamatea boundary; the musterers had just come in from mustering up there. 'He and his mate were fishing out by the Pinnacles, about where Peters meets the Swamp fence. It's a very swift-flowing river and rips the shingle out from under your feet. Good fishing there – he had some big fish and put them on a string, around his neck I think. He flipped over in the river and his mate could see him right down the bottom of this pool, held down by his fish.'[15] The rivers bounding and dissecting Ngamatea were good for fishing, yet there were remarkably few keen fishermen among the musterers and other staff. A few old hands like D'Arcy Fernandez still come back to fish at the Taruarau; Lance Kennett and his children fished down there when Lance was stock manager on Timahanga and Ray Birdsall kept the family and the cookhouse supplied with 8-pounders.

When a musterer did catch a fish, everyone remembered it. In about 1964 Park Pittar took a fly-rod out and caught 'an enormous' trout down from the Boyd hut. Park was very proud of that fish: 'I was just eighteen and it was a beautiful 9-pound rainbow – we weighed it – we had scales to weigh the pack-boxes. We cut it into big steaks and cooked it in the camp oven with onions. There were seven or eight in the gang and everyone had a really good feed of trout. We lived well that year.'[16]

One or two tried a nickel minnow – a bullet – over the years, but no one seemed to have any success, maybe because there were too many opinions about how it should be done. 'They say if you put your rifle under water so there's equal pressure all around, you can do it. We tried it in the Rangitikei – big hole and big trout – huge trout! – but I never shot anything – and I never tasted trout while I was there.' One said you could stun trout with a .303 off the swingbridge, and someone in the river with a fish net – a flour bag round a bit of No. 8 wire, to catch him before he recovered and swam off. 'You couldn't put a .303 barrel under water like a .22.'

Putting a bit of netting across the river and chasing a mob of packhorses down stream to drive the trout along didn't work: 'we never got a fish'. One time some of the boys tired of these performances and thought of a better way. The head shepherd got to hear of the plot and would have no part of it. All he would say was 'don't put your geli on the same horse as your dets'. There was an idea – why not pack both gelignite and detonators out on Tarzan: 'he was a sod of a horse and if the geli went off it might blow him to bits'. Now this is a true story; only the names have been changed to protect the guilty.

> Old Fred, our packman-cook, was a dab hand with geli so we got him to show us what to do. I don't know who the hell took the geli – but someone carted that and someone else carted the dets. Fred used to put about a quarter of a plug of geli and the fuse and the det and then he'd put some greaseproof paper round it and a stone. You'd drop the stone in above the trout. It wouldn't kill it – it would just stun him, and he'd come to the top and flap around. All these holes in the Ngaruroro where they were, were above rapids and we'd all spread out across the rapids, and this trout is flapping around and everyone's saying, 'There he is – there – quick – quick, quick . . . there – there.' One time Trev, one of the musterers, was giving Fred a hand. Trev used to ride a horse called Blaze. He was a good horse, but he couldn't swim. He'd fight in the water, his head would go under. So the trout's further out in this big hole and we let the geli off, and he was going to get away, and Trev jumped on old Blaze to ride him out to try to get him. He's flapping around and you had to get them fairly quick before they come to – they're just concussed. There's Blaze with his head under the water and Trev wants that trout and he's saying, 'Steady, Blaze, steady Blaze.' But Blaze couldn't hear because he had his ears under the water! We got our trout anyway – a monster, and Old Fred could really cook them in the camp oven. Delicious.

One of the best packman-cook stories comes from a later era. Peter van Dongen, who began life in the city, was not long out of college when he went to Ngamatea. He had had six months on the Rangitaik, working for Woody Collins, and through him he heard of Ngamatea and wrote to Lawrence, who offered him a job as shepherd-general, to start right away, in the middle of winter. But a couple of months later the packman-cook failed to turn up, so Peter was given the job.

> I'd never cooked anything, but funnily enough that didn't worry me particularly. . . . I didn't have any lessons from the station cook before we left either. When I asked him what I was to feed these guys on he just said, 'Chops, roast mutton and cold mutton.' You didn't really get help or advice from anyone. You were just expected to get on with it.
>
> It was all quite an adventure for a schoolboy to be involved in. We all saddled the packhorses and loaded the pack-boxes. They had to be weighted evenly. We didn't weigh everything – we judged it, but you got very good at it very quickly; you had to. I can remember the rations: there was lots of flour, spuds, tinned peas, we had tea, no coffee in those days, and milk powder and oatmeal and jam, cocoa, sugar. We made our own bread in the camp oven. We had raisins and sultanas, and I got quite good at camp-oven puddings.
>
> We set off for Golden Hills. It was a magnificent sight, it really was, and the Army was having big exercises out of Waiouru and they were flying quite low over us – big planes – and people were actually leaning out taking photos of this pack team.
>
> It was quite a logistic feat taking all the food for twelve or fourteen days for the musterers and the horses. Jack Aldridge, the station cook, would make up a couple of lovely big fruit cakes, and they would last for the first two or three days, and after they'd gone we were on our own. I'd get up about three o'clock, light the fire and boil the billy and get breakfast ready – porridge and chops. The musterers would be gone before first light, so they could get out and get onto the sheep before they moved. Once they'd gone I'd clean up, then I had a few hours to myself and I'd prepare the evening meal, and go and look for a deer to shoot or perhaps kill a bit of mutton. Then they'd start coming back about two or three o'clock in the afternoon and I always had the billy boiling and something to eat. The afternoon was quite relaxing, then we'd have the evening meal which was always a stew or a camp-oven roast, and a steamed pudding.
>
> I made one terrific gaffe, just after I began, and I kept it a secret for 40 years! I actually owned up and got it off my chest when we were at Peter Looker's sixtieth birthday party. I remember quite clearly we were at Golden Hills, probably my fifth or sixth morning there, and I thought I was getting into a pretty good routine. I always had billies of water, and kindling set aside from the night before to light the fire because I didn't want to disturb the musterers before they needed to get up, and I hopped out of my bunk, lit my fire, put the billy on to boil for the tea, and went about my chores. I was

working by single candlelight in that dark hut, the boys were stirring and coming to and if any hadn't woken up when I had the billy boiling I'd give them a shake and they'd have to get themselves organised in a hurry.

This particular morning everyone ate their breakfast and drank their big mug of tea and left on their mustering beat. By this time dawn was breaking and I set about cleaning up the breakfast things. I went outside and down towards the creek and I emptied this billy with the tea leaves – and to my absolute horror when I threw the tea leaves on the ground there was a pair of my socks amongst them! I knew straight away what I'd done. I'd put my socks to soak in a billy the night before – and I think I might have put a bit of soap in as well – and in the morning in the dark I picked this billy up and put it over the fire and boiled it and put the tea leaves in and served up this brew. I remembered then the tea did taste a bit funny, but no one had said a thing. They all had their porridge, had their chops, drank their mugs of tea and left. And I was mortified. I just sweated all day until those guys started arriving back. No one was sick, no one was off colour, there was the usual banter and carry on – and that was a secret I kept for all those years. Doesn't it say something for the constitution of back-country musterers![17]

People got lost at Ngamatea every now and then, often because they were caught out in the dark. Most of them smoked in those days which meant they always carried matches so if they were *really* lost they would light a bit of tussock, get some smoke up 'and the boys would come and find you'. One night Howard Sandilands, the packman-cook, 'an elderly joker, a panicky sort of a fellow', got lost in the midst of the muster and spent a night out in the open. He was a keen hunter and he had 'a real fancy rifle' so sometimes he would have dinner early and maybe one of the boys would wash up for him and off he would go for an evening shot. Ray Birdsall, who was a good tracker, was head musterer at the time.

We were at the Kaimanawa hut and he went out of the hut down to the Kaimanawa Creek, way down Forks Spur and he left it too late to come back. Next morning we had a three o'clock start, and no packman. We had to delay the muster for that day and people went looking for him. We thought he was over by the Rangitikei and by the afternoon we'd found his marks way down towards the river but we couldn't find anything else. About three o'clock I went higher, just in case he'd gone past – and there were his boot marks heading in the opposite direction. Instead of turning up to our horse paddock he'd gone right on past – up over the east–west watershed and on down towards the Ngaruroro. He was completely bamboozled. I raced back and got the boys and we took our guns and tore out on our horses – got right up on the tops – bang bang, a couple of shots and way in the distance we heard his answering shots. I've never seen a joker so pleased to see us when we arrived – we could see him way in the distance, running like mad towards us for a start. He'd slept up on top of the watershed, but he was good as gold – nothing broken or anything. He was just starting to head towards the station about 18 miles away because he'd come into view of it – he could just see the massive trees like little sticks on the horizon and if he'd gone down the gully he'd have been lost again before he climbed up out of it. But he said, 'No, no, I remembered everything they told me – I didn't have a drink of water – I sucked a stone.' Well, beautiful water everywhere and he was as dry as one thing. Someone had put him crook – told him to suck a stone![18]

A week or so later he'd got over his ordeal and thought he would go out for another shot. He was just about to go out the door

and the boys grabbed him and tied a string round his hand from the middle of a big ball we had and he says, 'What's going on? What the hell's this for?' And Morrie Mott says, 'Well, when you get to the end of that string Howard, start winding her up, and you won't get lost.' Poor old fella. At the end of the season he announced to the boys that he was thinking of finishing up. 'It's not a fair go,' he said. 'I wasn't really lost, you know. I knew where the station was, and I have to pay for you finding me', and he showed them his wages sheet – so much a week 'and found'.[19]

A few years later during the straggle muster, when the stags were roaring, the boys were camped at Peter's hut and Dave Withers and Dave Wedd were working back towards the Kaimanawa face when they heard stags down at the bottom of a gully. When they finished for the day they went out after the deer.

We blew a couple over and it was just on dark. Dave Wedd and I had got a bit carried away with these stags, and the night's clagging in on us, and it's blooming cold – you're up 4,000 or so feet, and trudging on towards Peter's Basin and where we believed the hut to be. Then drizzle, and fog, and it became pretty dangerous. We thought, We've got to make a decision here, we can't keep stumbling around like this – we were all on rock, no vegetation, just rock, on the top. We're only in Swanndris, long pants, boots, balaclavas – luckily; a bit of tobacco, wet matches – couldn't light a smoke. This is getting serious. We could tip over the edge. We've gotta camp here, so we found a little rock ledge to camp on – no cover, but for safety we had to stay put, huddled together. Nothing to eat: chewed our tobacco and spat and chewed. We camped on this rock ledge till the early hours, and we had our hands up inside our balaclavas, breathing on them, and every now and then we'd do a few gymnastic sort of exercises to stop us freezing up altogether. Daylight and we're out of there, over the ridge, down the face and into the hut – not far away. The boys were all there. And what did they say? 'Oh, you buggers come back? We've divided all your gear up, and the dogs, and your saddles.' They were relieved to see us come back, but they're not going to tell us that. Dave and I were blue – all day. The cold was deep into us. The boys gave us a hot breakfast, but we couldn't get warm. Great mates, good camaraderie – but gee, it was close. If it had snowed we'd have been in dire straits.[20]

Even in later days, with no real back country and bulldozed tracks everywhere, it was still possible to get lost, as Richard Whittington, the tractor driver, found out.

One night I was out there – way out, the last block before the boundary. I disced the track out and had my vehicle and diesel tanker parked there, and I was doing the eastern side, the downhill side, and always worked till nine or ten o'clock, and this one night fog came down, real thick. The John Deere's got a lot of lights on it but you could only just see to the end of the light beam, and I couldn't find my way back to the tanker. I thought I knew the block, but I was really only just into it and I hadn't memorised it yet. It was about 500 acres, no fences, no landmarks or anything and of course these little gullies come all the way up and rather than drive around every gully you think, Oh, I reckon I know where that one goes, I'll bypass that and go across the head of it. And same with the next one – but where was the vehicle? In the end I called up the cookhouse on the radio, and said to put my tea in the fridge, I'd just have to sleep out there that night, I was lost. And when the laughter stopped Tony Coutts who was possum hunting said he'd come out and guide me back to the vehicle.

A while later he came out and we both had CBs, so he parked on the track where my vehicle was, and he couldn't see me and I couldn't see him, but he directed me back by sound. He'd hear the roar of the tractor and he'd direct me right or left or straight ahead as the noise got louder or faded. It wasn't till I was virtually on top of him that we saw each other. The tractor was well lit up and he couldn't even see me coming out of the fog. And then, Murphy's Law, I was fuelling up and going to shut off for the night – and the fog lifted! I got a real ribbing about that for ever.[21]

Early one season during the main muster the whole gang was snowed in at the Log Cabin for several days and they were running out of mutton. Someone had to get out and find some, somewhere, so they drew straws for the privilege. Paul Thomas and Dave Withers drew the short straws and

Pohokura yards, branding a filly with original JF brand; Lawrence in foreground, Gordon Wright and Tommy Chase.

decided to head down towards the Taruarau with one packhorse to bring the meat home.

> We located some sheep way down through the mist – whipped a dog out around them and we had a mob of about 30 or 50, the dogs holding them on a flat area. You had to time your moment: one fellow holding them with the dog and the other has to whip in and catch and kill – in again and catch the next one. We need about six – two a side, and one or two over the top – they're not heavy sheep those merino crosses. We gutted them on the spot – no hydatids regulations in those days – but we'd skin them back at the hut – keeps them clean. Hygienic wrapping!
>
> We're right, mate, we got the meat, we're gonna survive, job's all done. Now . . . where are we? Cloud down to ground level, misty rain, and all the excitement – we're completely disorientated. Which way are we gonna go? 'I know the way – over there.' 'No – it's this way.' We ended up having a scrap, bloody near fisticuffs. But we had to make a decision, so one of us gave in and followed the other, and we picked up the track and got back to camp. We were that wet and frozen by the time we got back – no thermal gear in those days. I can still see Frank Brady and the boys standing outside the Log Cabin – drooling over that meat. But we had a helluva job trying to get the fire to burn up and warm us. And old Frank says, 'I know how to warm your hands

> up. I've been told – it's an old mountaineer's trick. We piddle on your hands and that will warm them up.' And you know we were quite happy for them to piddle on our bloody hands.[22]

The rest of the boys went to it with their knives and got the pelts off the sheep and took off enough meat for a feed – chops, fresh – into the camp oven.

Fresh mutton, 'almost kicking', was one thing, but the thought of rabbit going into the pot was too much for some of them. '*Underground mutton!* – No, oh no.' But Winnie Roberts used to make a wonderful rabbit pie; Jack drooled over the memory of it.

> She'd soak the rabbit overnight in salt water and then she'd make a thick white sauce and hard-boiled eggs and put it in a pie dish with a big thick pie crust over the top. Dad loved this, and he had the job of serving up the pie. There were usually extra people at Ngamatea, round the table, and Dad had the spoon and he tapped it on the table after everybody had had their first helping and when he had everybody's attention he said, 'Would anyone like more of the cat?' You can imagine the looks on the faces of those who weren't used to this – and he was sitting there with this big grin.

Bruce Atchison, too, recalled Lawrence's sense of humour.

> He used to cut all the young horses, the colts, up there, and it would take quite a while. We'd cut them and brand them, the fillies and all, with the old brand – JF – John Fernie – the original Fernie. And one day he said 'While we're at it we might as well do the pig'. So we catch the bloody pig, and we cut the pig, and we were coming back and there was a cat there, and he said 'We might as well do him too, he's a bloody nuisance'. So we grabbed the cat and shoved his head down a gumboot and castrated him at the same time.

When Lawrence went down to Marlborough or Otago to buy stud merino ewes and rams he often left the car at Napier and flew down. One time he was flying to Blenheim and Winnie suggested he send a telegram when he arrived, just to let her know they had landed safely. About the time of his expected arrival the phone rang and the operator said he had a telegram. Winnie asked him to wait while she got a pencil and paper but he laughed and told her she would not have to write this one down. It read 'Landed, Love, Lawrence'.

Another time Winnie and Lawrence were driving through to Otamoa to visit Walter and Eleanor Fernie and they came upon a large mob of sheep on the road. There was a car coming the opposite way and they met in the middle of this mob. Lawrence looked across, recognised the other driver and gave him a bit of a wave and a g'day. 'Who was that?' asked Winnie. 'Oh, that's So-and-So, he owns a farm back up the road a bit, and that's his wife.' 'His *wife*? She looks young enough to be his daughter.' 'Yes,' said Lawrence, 'that's his second helping of dessert.'[23]

Before there was ever an Army camp at Waiouru, the Fernies had the lease of a huge area of Crown land there. It was an important intermediate point between Ngamatea and the Parapara properties. There was only an old house there in those days, and a hut nearer the road for an old chap who worked for the Fernies.

> Wethers would be spread out across that country and they could be fattened there and railed to Wanganui. What was left after the pick was taken out were walked down to Pukeroa, near Hunterville, and they would be replenished from Ngamatea. Pukeroa sent their fat wethers by truck to Wanganui and what was left over were walked to Otamoa. Then fats were taken out of those and trucked to Wanganui and what was left walked to my brother John at Rusthall to be fattened, and from there they went to the works. That was their thing – from Ngamatea the Fernies more or less walked their sheep to the works. Pretty old sheep, but they were in great order – marvellous teeth.[24]

A couple of musterers would stay on at the end of the season to drive a mob of sheep or cattle between Fernie properties.

> From Ngamatea we'd take them as far as the river. The people who worked for Otupae would give us paddocks to put the sheep in. Then up the Rangitikei Hill to Ohinewairua, stay the night there, down to Moawhango, to Te Moehau – Batley's place, from there wander out to the main road, as it was then before the deviation. That was our biggest day, driving them from Te Moehau holding paddock right through to Hihitahi – but you didn't get the lunatics on the road that you get now. They respected stock and dogs. Very little traffic. Then you followed the main road right through to Waiouru.[25]

Several of the boys had twenty-first birthdays while they were at Ngamatea. Bill Cummings happened to be camped at the Log Cabin by himself on his big day: 'I had a drink of cold water!' In Ray Birdsall's first year he was 'mustering round the back of the Otupae range and getting the sheep up through the fern. It was my twenty-first birthday. I never said a word to anybody and no one knew. And it was one of my worst days mustering. Nothing would go right – the sheep were weak and the fern too big.' Jack Roberts's twenty-first birthday party was held in the Timahanga woolshed. Everyone on the station was invited; Bruce Atchison remembered Jack bought two 18-gallons and the boss said, 'Are you gonna float the bloody *Titanic*?'

Dave Withers was keen to work at Ngamatea. He had been at Smedley, a training farm in central Hawke's Bay, and it had given him a taste for big country. At the Smedley dog trials he met Bob Passmore, a great dog-trials man, who had been at Ngamatea from October 1955 for a season and a bit, and when Dave told him he was hoping to go there Bob said he was too young – the boss wouldn't take him until he was 21.

> I thought about this for a while, and my yearning to be up there and be part of it was just so strong I wrote to Mr Roberts and told him I was 21. He accepted me – and I was nineteen coming up 20. The next season I turned 21 and we were shearing down at Timahanga. My job for my twenty-first birthday day was a 3 a.m. start and take a mob from Timahanga back to Pohokura and put them across the swing bridge over the Taruarau. Al Cooksley, the bookkeeper up at Ngamatea, was doing the mail run this day and I'd organised with him for two dozen beer to be delivered to the quarters at Timahanga so I could shout for the boys.
>
> When I got back up to Timahanga I went over to the stables with my horse and was walking back to the woolshed when I saw Lawrence coming down from the shed. He beckoned me over. He seldom called anyone by their first name – always the surname – and he called out, 'Wwithers, come here.' I thought, Gee whizz, something's wrong here and I went over to him. He looked me straight in the eye with those steely grey eyes of his, a bit of a smirk on his face, and he stuttered away there. 'Wwwithers, I've got a mmessage for you – from your mmother.' 'Yes. Mr Roberts,' and I started to blush and felt myself shrinking, knowing what was coming. 'She says Hhappy twenty-first bbirthday.' And this big grin comes across his face and he put out his hand and shook my hand and without another word turned and walked away.[26]

Paul Thomas spent his twenty-first at Pohokura at the end of September 1959. There was usually one shepherd stationed down there with old Leo McSweeney, the Pohokura cook, and Paul had been there for four months and did not join the mustering team until the middle of November. Leo happened to be away that week so Paul was on his own. But his birthday was a Friday, mail day, and he would always ride or drive up to Timahanga and, for a change of company, have a cup of tea with Mary Chase, the Timahanga cook. But she was not there that day either.

> I had to ride up because the track was too wet at the time and I missed the mailman, so I picked up the stores and put them on the horse and rode

back down to Pohokura. It was quite dark by the time I got down there, and it was drizzling and cold. I tried to get the wood range to go again, but I gave up in the end, so it was a glass of water and a piece of fruit cake for my twenty-first birthday, and I didn't see anyone else all day long. My dear old aunt had sent me up a big fruit cake – it was in the mail and it was on the front of the horse as I went back down. And a very heavy pair of gold cuff-links too, I might add. Just what I needed at Pohokura! I think there have been better ways of spending a twenty-first birthday – but I was happy. And Dave Wedd and Dave Withers were up at Ngamatea and they both had twentieth birthdays around that time and I went up to the station one Saturday evening, just before the mustering season began, and we celebrated the three birthdays.[27]

When all the musterers were there they would occasionally have a bit of a party: it was a change from life out the back. There was room to move, room to sit around and have a few beers, time to have a good yarn. If Ted Craven was on the station they would ring him up at the homestead and he would come down to the quarters and join in the fun. 'He used to play one of these old button accordions, as old as billy-o, and we'd all have a bit of a sing-song. It was really good.' And there was always the dance at the woolshed when shearing cut out, and maybe another during the year if there was any good excuse. Two of the deer cullers who were working out at Golden Hills, walked in to one dance. 'They were camped out there. They had no good clothes, just one pair of sports trousers, and one of them put the sports trousers on and he'd have a dance and straight after that they'd go behind the wool bales and the other joker would whip off his shorts and put the sports trousers on and then he'd have a dance.'[28]

By the 1960s there were more cars around and the boys could get into Taihape occasionally on a Friday evening or if they were lucky someone would have toothache and they would have to take him to the dentist. They would get into all sorts of mischief in town – 'just good harmless fun'. Their shopping done, they would try to be at the Gretna before six o'clock closing, and the good burghers of Taihape, the police included, were extraordinarily charitable to the boys from Ngamatea. They might be looking a bit rough and not entirely sober, but they were looked after. They could book up clothing and other gear to the station – it might take a phone call, but as often as not a signature on the docket was enough. Everyone in town knew Ngamatea would honour the dockets and pay the bills, and the station simply recorded the debits in the wages book.

The cinema was the main entertainment back then. Forty years later Jack Roberts and Dave Wedd hooted with laughter remembering their antics. 'One evening,' said Dave, 'the place was nearly full and the usher took us in and put us right up near the top. As we were going past it was too big an opportunity with the light coming from the projector out onto the screen and old Frank makes an obscene gesture with his hands. Well, they were up there in a flash, and they had us over to the police station – just bad luck the police station was right opposite the cinema. The cop on duty asked us where we were from – and he used to come over hunting at the Tikitik, so he wasn't going to do us any harm. He just sat us down by the fire and he said 'You boys better stay here till you sober up.' Then after about half an hour he said, 'You're all right now' and he took us back over and made the usher put us back in our seats! Yeah – it was all a long time ago . . . You wouldn't get away with it today.

One night Jack went to Taihape with the boys for

> some picture that was worth going a long way to see. The theatre was jammed full, cars all up and down the street, and when it was over Frank gets out in the middle of the main street in Taihape at the corner of the street where the cinema is, and he starts directing the traffic. In about five minutes he had the biggest bloody tangle up you've ever seen. Everyone's tooting on their horns and there was a real uproar. It was only about 100 metres up the street to the cop shop, and the cop came down and grabbed Frank and I stepped forward and said to him, 'He's with us', and the cop says, 'And where do you come from?' 'Ngamatea' – and he just said, 'I'll give you two minutes to get out of town.'

Dave also recalled the pub raffles at the Gretna – 'they were always raffling something – a bit of wild pork . . . Anyway this night it was a muscovy duck and they had it in a sugar bag – just sort of showed you the bag, and everyone buying tickets, and one of the Ngamatea boys won it – and it was a seagull!'

On another occasion there was a circus in town. The boys did not even get to see it – they had had too good a time at the Gretna – but somehow they got caught up with some of the circus people.

> I remember seeing these elephants, and we shouldn't have been in the condition we were, with these great big brutes. 'Stick around,' they said, 'we'll be loading them on the train at such-and-such a time, and we've got to take the elephants and the horses through town.' So we hung round. It was dark and we had to go from the circus once it was finished, up from the showgrounds which are down towards the river, and right through town and over to the railway station.
>
> Well, we rode these circus horses through town – Withers, Frank Brady and the rest of them – these beautiful palomino horses women used to

ride and do all these tricks. Someone else was helping with these two great big elephants, walking through town – and someone broke down in the main street and the elephant towed them to get his car started. The joker just backed him up and pulled this car with a great big rope! We got to the trucking yards there and they just opened these doors, and in they went, onto the train. Commonplace with cattle and everything back then, huge mobs. But those elephants are very big animals when you're right beside them.[29]

When Frank was not playing up in town he was busy looking after Ngamatea horses. The winter was pretty hard on them and each year when the season started it would take a while for them to get back into condition, so Frank got the idea of taking a couple of them down country for the winter. He used to do a lambing beat at Olrig station, down the road from Big Hill and Kereru, and he would ride one horse and lead another from Pohokura up and over the range and down to Olrig. The horses got hard feed through the winter and went back for the next mustering season in good order.

I'd spend the night at No Man's, up on top of the Ruahines. It's not very far from there down to here at Olrig – a few hours. I could easily have come over in a day – no problem. Years ago, before Fernies had Timahanga, Jimmy Leonard's sheep used to be brought from there up over the top and down to Big Hill for shearing – they'd drive them down, and the story goes that on the way back they stopped at Ruahine hut – that's about halfway up to No Man's. They had these sheep in a makeshift yard they'd set up in the bush, but there was a terrific snowstorm and they lost three-quarters of them that night. Those were the things that finished it for them, farming on Timahanga.[30]

At some stage early in the history of the Owhaoko block a surprising amount of fencing was done out in the back country – the Golden Hills, Gold Creek areas and also across the river on the Harkness block. The musterers would come across the occasional wooden post and rails, and also wire, and metal posts and standards. There were no standing fences, but some of the wire, especially over on the Harkness, was still serviceable and Ashley Watt thought he had a use for it on his father's dairy farm at Nuhaka. As far as Lawrence was concerned, the place would be better without it and he said Ash could have the wire if he cared to go and roll it up. Don Hammond went back for a couple of weeks in the winter to give him a hand. They took a packhorse, stayed in the Boyd hut and packed coils of wire back from the Harkness to the station. 'And it was *cold*,' said Don. 'We had the fire going in the Boyd hut all night. I remember one night in particular – we had a big tin pan we used to wash in, about a foot deep, and this thing froze solid inside the hut.'

Dave Wedd and Dave Withers copped another winter fencing job when Lawrence and Winnie went away for three or four weeks, possibly to Australia. They were instructed 'to do up' the old rabbit fence across the back of the Swamp.

> It was the only fence out there. You started at the Taruarau, above the Woolwash. We took a packhorse each day, and our lunch, and we'd ride out – and then when we got to the far end, over towards the Rangitikei, we'd stay in Peter's hut and work from there. We couldn't believe we were trying to fix this fence, just the two of us. We took our time. For two or 300 yards there might be about three wires, and no battens, maybe a few steel standards. That was all – the netting had collapsed and was all overgrown. For a stretch it wouldn't be too bad, then it's through birch bush, fallen logs, hard to find the fence. When the boss came back he said, 'Did you fix that fence?' and we said, 'Oh yes, Mr Roberts.' 'Did you see that spreader?' he said, 'up on top of the Taruarau there?' 'The spreader? – ah . . .' 'Well, you better go back and fix that.' We hadn't seen no spreader – and that was the beginning of the fence – a great big post they thread the wires through! He just knew how to catch us every time. And we were so proud of what we'd done.[31]

When Ray Birdsall came back to Ngamatea as manager, he had quite a cooperative relationship with an SAS unit from Auckland, who were looking for a place where they could train in the bush: Golden Hills and the Peter's area were just what they needed. They also wanted somewhere quiet and secluded where the troops could sit their exams, and the shearers' quarters suited them very nicely. In return they would cart fencing gear out in an Iroquois helicopter to whatever fence line was under construction at the time. The first time they visited the station Ray was surprised to see this big chopper drop down near the homestead and a dozen or so men hop out of it. When Maxie appeared the major explained that they 'happened to be in the area and thought they would drop in and see what sort of a place it was'. Maxie, of course, invited them in for a cup of tea, and when just two officers stepped forward she said, 'Oh no, all the men can come too.' They did, Ray came up from the yards to join them, and they soon came to an agreement. 'You'd get up at three o'clock in the morning,' as Ray recalled, 'and go to catch your horse – right beside the shearers' quarters – and there's a guard outside the door, with a rifle and bayonet fixed and one of the officers would see me and he'd race out and say, "Come and have a whisky." That's all I needed at that hour, but they were relaxing after exams. They were real dags.'

One year when the troops went out to Golden Hills for some training they left some of their new gear in the shearers' quarters, but did not think to tell Ray that and took off for Papakura before he saw them again. A week or two later he got a letter from the major to say that all the new clothing they had left in the shearers' quarters had disappeared when they came back looking for it, and he asked if Ray could find out what had happened to it.

> Well, I was up in the blacksmith's shop a few days later and in walk these two Rabbit Board rabbiters we had on the place at the time, terrible people. It's first thing in the morning and they're in brand-new khaki shirts, new pants, new boots – Army gear. Away they went rabbiting. I didn't say a word, but I had a friend in the police in Taihape, fairly well up, plain clothes – and he knew the Army were out there. Rang him and told him all this gear had gone missing and this morning two rabbiters had arrived all dressed up in new gear. 'Right,' said Bert, 'I'll fix them. There'll be an Iroquois over you in no time; I'll get the military police from Waiouru to bring me out.' So out they came and they landed by the stock yards, and said 'Where are these chaps?' I said they were rabbiting in the Gully block and Bert and the MPs said, 'We're not waiting for them to come back – we'll go and get them' – and up in the helicopter. Landed right beside them, on the job. Here they are all dressed up in their new gear. Well, they stripped them right down to their underpants, and these jokers had to come back like that – no boots, no socks, nothing. It was the joke of the farm for ages. Came back in their underpants! They should have taken them too.

Ngamatea stories were passed down from generation to generation, and everyone felt they knew the place and that it was part of their heritage. The lucky ones got to visit up there, either before or after they worked there themselves. As a small child, Lance Kennett used to go up when his father, Alan, went back each season to shoe horses. His mother would take Lance along for the day when she drove Alan up, and when she went back to get him a few weeks later. The really lucky ones holidayed up there. Shorty Chapman and Ken Fraser, and sometimes Jack Cornwall, took their families up two or three times when Ray was managing the station and camped in the shearers' quarters over Easter or a long weekend.

The kids wanted to live rough as their fathers had done – use a long-drop, wash in the creek, eat food cooked over the fire. They would not take any toys, they would not eat the dinners their mothers cooked on the stove, they carted milk and butter over to the Woolwash creek to keep it cool

The Tikitik hut in winter 1959.

– 'Dad didn't have a fridge'. Each of the kids had an ice-cream container, a plate, knife, fork, spoon, mug: 'that was all Dad had when he was a musterer'. They invented games, made bats out of bits of wood and played rounders. And, as Ivan and Elaine Chapman remembered, they loved it.

> We took a TV with us once for some special event that was coming up – an old black and white. Ray and Maxie had very poor reception up there, you could hardly see it at their house, but we rigged up an old aerial the shepherds had had. It was blown over in a paddock and Ken and I put a post in, just outside the shearers' quarters, hoping we might get a picture. Elaine and Pet, Ken's wife, were inside, looking out the window and we were reaching up trying to turn this thing round with a spade and the girls would call out, 'Right – that's good – hold it at that', so we'd tie it off, put the spade down ... and they'd say, 'Oh, you've lost it – it's gone off again.' And we'd never moved it. We couldn't make it out till Ken hung the spade over the thing – and all of a sudden we clicked. It was the spade that was doing it, so we left it there.
>
> The kids weren't interested – they were doing things with the fire – but Ray and Maxie came down that night and they could not believe the reception we were getting. They had a huge great big aerial with lots of bars and things stuck way up in a tree and all they could see was red and blue, and here we were down there with a spade for an aerial and it was clear as a bell.

And Ivan took the kids out to show them the old Tikitik hut, and where the horse paddock was, and where they used to tie their dogs. They'd seen plenty of photos of huts, but they wanted to *see* one, and this was the nearest – we could drive part way and then walk. It had just about fallen down; they were shepherding by this time, not mustering and they didn't camp out any more. But the kids were so funny – all those years later they were rushing around hoping to find the very things Dad used when he was out there.

It was Ray's birthday while they were there so Elaine and Pet did some baking: the children liked these cakes at home, but because musterers never had them, they did not want them either. There was even a birthday cake, but 'we had no colouring to write his name, so we got some beetroot juice and dripped it into the icing we'd borrowed off Maxie and iced "Happy Birthday Ray" in pink! They came down and ate with us that night – and he thought that was great. And the kids thought this was the *best* holiday they'd ever ever had.'[32]

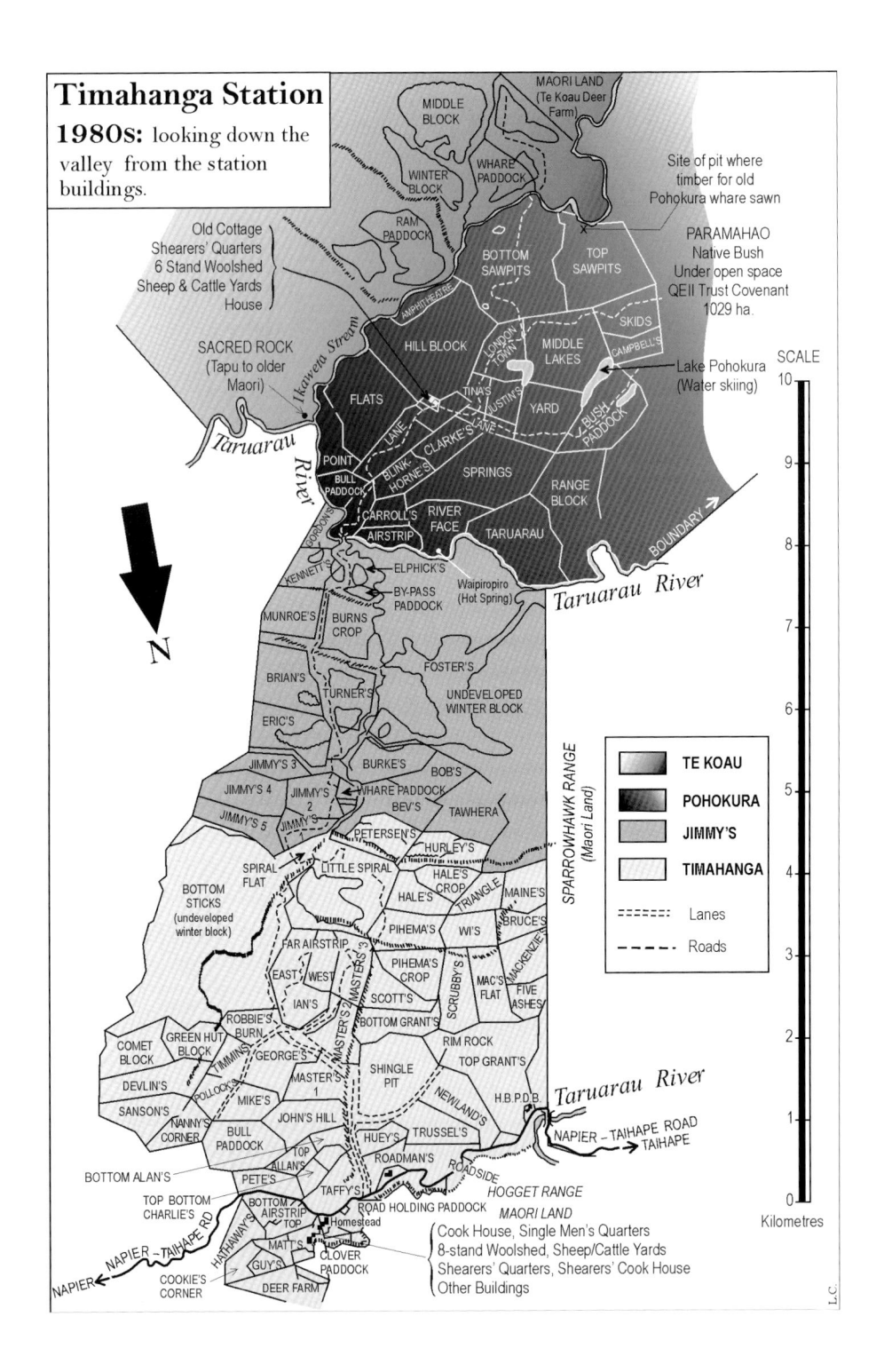

MAP 4: *Timahanga Station*

Chapter Six

Jack and Timahanga

*Given what they started with, developing such a property
as a family enterprise is truly outstanding.*

When Jack left Ngamatea in 1951 for Napier Boys' High School, his father's old secondary school, he found it all a bit traumatic. He had spent very little time away from the station, and none away from his parents or close relatives. Lawrence and Winnie drove him down to Hastings and they all stayed overnight with their friends the Wellwoods. About an hour before they were to take him to the school hostel and launch him on the world, Lawrence suggested he and Jack go for a little walk. The Wellwoods' house was set back from the road and they walked in silence right out to the end of the long drive where Lawrence stopped, turned to his son and said, 'Remember, whatever they do to you, they can't eat you.' They returned to the house without another word.

For the first term parents were obliged to keep their distance lest they unsettle their offspring. There were plenty of other country boys, from Wairoa down to the Wairarapa, and there was one neighbour boy, Hamish Lyall, from Otupae, but he was a fifth or sixth former, another species to a new boy. There was an agricultural course at the school but Lawrence and Winnie wanted Jack to get a good general grounding before he made up his mind what direction he might take, so they enrolled him in a course that included Latin and French. He dutifully sat through that for his first year, then changed to a more practical course, including woodwork and maths, which suited him better and which he felt would be more useful.

The school had its own farm and Lawrence donated a few halfbred sheep to give the agricultural students an idea of the difference between breeds. Barry Roberts brought them down on the back of a truck in a crate made from some old gates, and they were turned loose on the school farm – never to be seen again. The yards and fences were built for short-legged Romneys, not long-legged halfbreds; they just hopped over the fences and were gone.

School improved after a year or two, and Jack settled well. In his last year he was a school prefect and head prefect of the boarders, who were 'regarded as the cream of the school. They were on view most of the time. If you were a boarder you had to have your socks pulled up and your shoes shined and be neatly dressed. Look out if they caught you up town doing anything that would reflect badly on the school.'[1]

When Jack came back to the station after five years at school, Joe Larrington had built a new two-vehicle garage with a sleepout right down one side with two bunks and two big windows. 'That was my room – a beautiful room, really warm. It got all the sun, right round, and I lived in that for the next few years.' He drove the station truck for a start, then tractors, and seemed happy working with machinery rather than stock, but at the beginning of 1959, when they were short of a packman-cook, the boys asked Lawrence if they could take Jack out with them. It would do him good, they reckoned. Jack had done no more cooking than the next fellow – and less probably with a mother who cooked as well as Winnie – but he 'was damn good at it – he took it on and away he went. I think that was the turning point for Jack from being interested in machinery to going the other way. I remember he said he used to think we were a pack of moaning buggers and now he could see we had plenty to bloody moan about at times. Next thing he's off down the South Island and he never looked back from that point on.'[2]

Jack left in August 1959 and was away three years, mustering on high country properties. He started at Grampians, then went on to Blue Mountain. 'While I was there I went and did the autumn muster on other places – Clent Hills, the boss's brother's place, and Clayton station, a bit nearer Fairlie.' During a stint in the Mackenzie country, at Haldon, he also went to Lillybank for the autumn muster.

He was back at Ngamatea for the 1962–3 season, a fraught one as it turned out. Lawrence wanted his son to be second musterer, but Jack wanted to be 'just one of the boys'.

> I didn't know the beats. How could I tell people what to do when I didn't know where I was going myself? But Dad insisted, and the head shepherd who had been there two or three years got disgruntled and he and two of his mates decided they were worth more than the rest of the team and if they didn't get it they'd walk out. We were doing the Kaimanawa, halfway through the first muster and they were the only ones who knew the country – the rest of the gang were new that season.

When they came in and put an ultimatum to the boss, Lawrence, in his usual way just said to Alan Bond, 'Well, you've got a job to do, boy. You'd better go and make up their wages because they won't be here tomorrow.' They were down the road, and the boss went out and gave a hand with the muster while he found replacements for them. He turned to his loyal men from past seasons and Dave Wedd was able to come back at short notice; he and Arthur McRae arrived right after Christmas. 'We headed out to do Boyd's and the Right Hand Fork – we weren't going as far as the Harkness then, and we did up towards the Mangamingi and Boyd Top. Full gang – Dave, Arthur, John Smith, Peter Looker, Bill Summerton, Murray Taylor, and me – and only Dave knew the country.' When Dave and Arthur finished up Park Pittar and Kevin Sayers arrived and the gang settled down for the next few seasons. Dave thought the walkout unfortunate, a sign of a new attitude to the job. 'In our time we had men of experience that had four or five – Bruce ten – years, and they told you, Stick to the job or you go. You don't go to the boss saying you want more than the others. He had his way of running it, and you knew damn well if you didn't measure up, you went.'[3]

At the end of the season Jack went down south again to do a wool course at Lincoln College. 'It was a three months' course, but to get my ticket I had to either class a clip in a shed, or do so much time in a woolstore, and I just didn't have time to do the practical.' It was the end of Jack's apprenticeship; he came back as head shepherd for a season, then took over the management of Timahanga.

In June 1964 Jack and Jenny Paton were married. Jenny's parents, John and Norah Paton, farmed further down the Hastings road, near Otamauri, where they ran a Corriedale stud. Jenny was at school with Margaret, and did a Diploma in Horticulture at Massey while Margaret was there, and she was one of the bridesmaids when Margaret married Terry Apatu in December 1962. Margaret and Terry had met at Massey when Terry was doing part of his Rural Field Cadetship there. Terry's parents, Wilson and Molly Apatu, lived at Mangateretere and had other properties around Hastings, but after their marriage Margaret and Terry lived in North Auckland for several years.

It was the beginning of a new era at Ngamatea. Lawrence's health was deteriorating but Uncle Walter was still active so Jack had no freedom to be innovative. He and Jenny had about six weeks at Ngamatea before their new house, the first homestead ever built at Timahanga, was ready. There was little else down there apart from the woolshed, cookhouse and shearers' quarters, and the stables. There were no shepherds' quarters: the musterers lived in the shearers' quarters when the shearers were not there, and when they were, they shifted out into tents.

> When I worked down here at Timahanga when I was single, I stayed in a tent too. There was an old hut that our eldest son Johnny's now got down at Pohokura as a schoolroom, that Dad and the wool classer used to sleep in,

Schoolroom at Pohokura, formerly the old hut at Timahanga, which the wool classer slept in.

and there was an old malthoid hut outside the cookhouse, but that burned down later. It had an open fire and the boys shovelled the wood on it and it got that hot, up she went. When we came we got a new cookhouse, the boys' quarters, the rec room, and with some of the timber we got out of the bush up here we built the carport down at the cookhouse, our carport out here, and then the new stables. The old ones are still there, but quite unsafe now – the loft anyway. They're due for demolition, sadly. We've given it a coat of paint to preserve it a bit – at Jenny's insistence. I was all for pulling it down. The shearers' cookhouse is due for demolition too.[4]

Jenny was a good country lass, her father's offsider when she was little, 'helping' with the sheep. She suffered badly from hay fever, and it was only when she lived at Timahanga that she realised horses were the cause. She chose to do horticulture at Massey 'with a view to getting work afterwards' but there was no garden at Timahanga, just a house set in a paddock – and that became the lawn. 'It was about 2 feet high and we thought we'd be smart and borrowed one of those forage harvesters that squirts the grass out. It was blowing from the west and we thought we could mow it and it would all blow away – but the wind coming over the trees made a downdraught and it all ended up back in the carport!' The nearest neighbour was

The old Timahanga stables.

the roadman, when there was one. Ngamatea was eleven miles away, and it was about two hours to Hastings and one and a half to Taihape. Jenny had met some of the musterers when she and Jack were engaged, and 'the whole Ngamatea crew who were there that season came to the wedding'.

On her first visit to Timahanga, about a year before they were married, Jack took Jenny and her parents down to Pohokura. At Jimmy's there was a sort of S-bend, but there was a short cut, straight over the bank and down to the bottom. 'Of course Jack took it and nearly scared us silly. We were watching the road and suddenly we just went over the bank! That was one of my first major memories. You didn't have to follow Jimmy's Creek – the road had been put through at that stage but it wasn't reliable: sometimes you could get down, sometimes you couldn't.' There was still no bridge over the Taruarau, and they had to ford the river to get to Pohokura.

Winters were harder at Timahanga than Jenny was used to, but not as bad as those at Ngamatea.

> They're 3,200 feet up there, and we're only 2,300 feet. We had no power – it was all generator, and we only had the phone to the station and had to relay messages through them. It didn't always work too well: you only had to leave out a 'not' to change the message entirely. We had a radio-telephone in our station wagon so we could get Overland Transport at Fernhill or the depots in Taihape. They were on the air at 7 a.m. and 7 p.m. – and Manapouri tugboats

and a cartage firm at Whangarei used to come in loud and clear! There were some very interesting exchanges on the air. We didn't get an outside phone for some years, and then we were on a party line of course with all the properties down to the Rangitikei River – ten people I think on the line, and there was only one phone on Ngamatea.

I was busy trying to make a garden but all we had around the house was some netting and standards, and when the sheep got hungry they came through, and the horses came through. And we used to have single men cooks down here, and half the time I ended up being the station cook. In our first year we either had eight cooks in eleven months, or eleven cooks in eight months![5]

They had a 'standing order' for cooks with Overland Transport, and they arrived one day with a new cook who was 'absolutely paralytic drunk'. Jack and the boys got him to the quarters and into his bed and left him to sleep it off, but when Jack shook him awake in the morning he asked blankly 'Where am I? What am I doing here?' When he heard this was Timahanga station and he was the cook he said, 'Like bloody hell I am', and staggered up to the road to catch the first vehicle that would take him back to the Bay.

Johnny, the first of their three sons, was born in 1965 and was just over a year old when his grandfather died. Lawrence had been taken to hospital in Taihape at Labour Weekend 1965, was back home for a while, then he and Winnie had to move down to a flat in Hastings so that he could have medical assistance and treatment. Towards the end, when Lawrence was back in hospital, Jack and Jenny would take Johnny to see him. 'He'd sit on the bed and you'd see Dad's hand sneak out from under the bedclothes and hold his foot.' Friends came from near and far to visit him, and his ex-musterers would call in if they could. Bruce Atchison was not far out of town, on the Taihape road, and he used to go and visit Lawrence and amuse him with his stories and cheerful humour. Alan Bond, who had been Lawrence's bookkeeper and handyman for a couple of years from September 1962, went to the hospital just days before Lawrence died: 'Jack took me round to see him. He was just a shadow. It broke me up.'

Lawrence died on 25 October 1966. At Jack's request Lawrence's head musterers carried him into the church for the service; his family carried him out. It was a huge funeral. The boys from Ngamatea came from wherever they were to honour their old boss and grieve with his family. Winnie knew how much they had respected him and how loyal they were, and they kept in touch with her and visited her at her Te Awanga house until her own death in 1980. She was never without visitors, never without memories of Ngamatea all around her. Norah and John Paton also moved to a house

Looking across Timahanga woolshed to Boyd's, 1960.

at Te Awanga, through the back fence from Winnie, and Ted and Joyce Craven were just a few houses further up. Winnie was thankful for all of this, but she longed for Ngamatea.

While Lawrence had been sick Jack 'sort of ran the whole place for about a year', then just before their father died Jack and Margaret asked Ray Birdsall if he would come back as manager. It was agreed he would manage Ngamatea and Jack, Timahanga, although the whole property would still be run as one. 'I was supposedly manager of the Timahanga–Pohokura block, but I didn't have the say in what should be done. I was the boy really, paid wages. We were governed by Uncle Walter – he was the king pin and we did what we were told. I couldn't decide on development or anything.'[6] It was not until 1972 that Margaret and Jack as part-owners could make separate agreements with Walter Fernie regarding Owhaoko–Ngamatea and Owhaoko–Timahanga and run the place as two separate stations. Walter died in February 1975 at the age of 97; his wife Eleanor died eighteen months later, leading to the third lot of death duties in ten years. The station accounts were in a fine mess. The old family accountant 'had let things slip and was about seven years behind doing the books'. He got the accounts up-to-date when Lawrence died, and that helped; there was giant discing and other development that could be set off against the first lot of death duties.

Jack now owned a property of about 26,500 acres of which Te Koau and Pohokura were freehold, and Timahanga pastoral lease, 'but of course the freehold properties were right down the bottom end, and when we started developing we didn't know whether we could get the freehold up here'. In 1964 there was not even a road fence; there were just three blocks, partly fenced, and the cattle roamed freely in the scrub.

> From the Taruarau River at Pohokura there was just a chain-wide track cut through the second-growth scrub right up to Timahanga to bring the sheep up. We started up here with those rotary slashers, put up a lot of netting fences and put in the airstrip and the super bin in 1966 and threw the fertiliser on. Prior to that, in the fifties, Dad used to get the Lodestar and then the DC3s direct from Napier airport. Some of Ngamatea was done too; when I was mustering in the early sixties the Lake block was burnt and at my insistence Dad got the big plane and we oversowed it with clover and ryegrass and supered it.
>
> A lot of the work that had been done down here in the forties and again in the fifties was reverting back into scrub when we came in the sixties. There was no fencing, and not enough stock. After 1972 when we split the two stations we were able to get going and we redeveloped everything – much of it for the third time. Dad did the Springs block when I was at boarding school and we did it again in 1980, and the scrub was massive. First thing we did in 1972 was put up a lot of netting fences to subdivide those blocks – we're replacing some of it now. It's served its time, but it's done well. We built a lot of hay barns scattered around the place relatively early on, although we didn't make hay to start with. We only started off with about 150 cows, down at Pohokura, and a few steers up the top here. We just built our numbers up from that. Some of the cows were fourteen and fifteen years old, but so long as they had a calf we tried to keep them going as long as we could.
>
> At the division, everything was divided in half – the stock as well as the land. Pohokura ran up over the Otupae Range and out to the mail-box and we just took the top of the range as the dividing line. Margaret and Terry owned Kaimoko – Boyd's bush – up here, and I took that and the whole valley here in exchange for my interests in Ngamatea. We were scratching for enough feed at the start, but it wasn't such a huge increase in stock numbers – they weren't that well stocked at Ngamatea. For a while we concentrated on Timahanga to try to get the land that was reverting into production, then later down at Jimmy's and at Pohokura. Hard to say what acreage, but now there's about 11,000 acres between Timahanga and Pohokura that's developed.[7]

Development of the valley depended on getting bridges across the Taruarau and some of the streams. The only bridge was the old one-lane swingbridge which all the sheep had to cross when they came up from Pohokura to be shorn at Timahanga. It was a major obstacle, especially

Gordon Grant on a Land Rover fording the Taruarau underneath the swing bridge.

when the musterers were trying to bring ewes and lambs up. Such vehicles as could negotiate the valley had to ford the river. If it was high they would slacken off the generator and fan belt, pack sacks over the front of the radiator and over the top of the motor, then take their chance in the current. A road and bridge was one of Jack's earliest projects, even before the properties were split. 'We put the road down the little Spiral when we took the timber out of the Paramahao bush at Pohokura. We got enough timber to pay for the road and then we stopped milling. There's still thousands of feet of timber left in that bush, but we've put it into the Queen Elizabeth II Trust to preserve it. We put the Kaimoko bush and the rest of that block into the trust as well, so we've got over 4,500 hectares in the trust now.'

The first bridge, built by the logging contractor, went when a big flood took the abutments, and it was replaced by a steel bridge with a pier in the middle. That was later strengthened with extra stringers to take the weight of stock trucks and trailers. For a while Pohokura was the only place clear enough to run sheep, but later, because they concentrated on clearing the Timahanga end, Pohokura went back into second growth and then they did a lot of root-raking and clearing down there as well. When that method fell out of favour because of the amount of topsoil it displaced they used brush-cutters – something like a weed-eater with a saw blade that will cut scrub 4 to 6 inches thick 'if you nick 'em both sides'. They used these first

View down the valley towards Pohokura, from Jack and Jenny's garden.

on the Springs block from the river right up to the bush; the next year the Taruarau and the Range block at Pohokura; then down through the Burns, between Jimmy's and Pohokura. Jack had taken on a massive project.

Fencing was the next priority, miles and miles of it. As the scrub was cleared and burnt it was oversown and fenced, and dams and tracks were put in. 'We oversow by helicopter, with a spinning bucket – it's more accurate. Then we super it with the plane off our strip here. They can take nearly 2 ton at a time in perfect conditions.' The airstrip is on a gentle slope and drops away at the end into the hay paddock across the road. When it first went in the planes were too close to the top of stock trucks and wool trucks and they had to change the angle of the strip slightly so the planes were higher over the road. 'We also put signs on the road – aerial top dressing, duck your head. We know of terrified travellers who nearly went off the road there when a plane roared over their heads.'[8]

These were the years of 'land development encouragement loans and supplementary payments on wool and all that, but we stopped all development in 1984 when there was a change of government and they just put the axe through the lot. Probably a good thing in some ways because it gave you false ideas, and a lot of land was developed around the country that shouldn't have been developed.' They were still paying for Timahanga on a deferred payment licence from when they freeholded the pastoral lease on Pastoral Run 23 in 1964. 'It took years to pay off, but it's been paid for quite a few years now and the whole property's freehold and we don't owe anybody anything on it.'⁹ These days from the homestead at Timahanga you look out over rolling green paddocks, all the way down the valley to the end of the property 16 kilometres away at the foot of the Ruahines. It is a far cry from the endless scrub Jenny used to look out on, and now she and Jack have two sons and their wives and five grandchildren living down the valley, where once there was only ever a cook and a single shepherd.

Jenny taught her three boys at home by correspondence. 'It's hard work – you have to be very dedicated, because there's always something going on here – the phone rings or visitors come in. We had the schoolroom down the far end of the house and the phone would go and I'd have to come and answer it, and by the time I got back there was nobody left doing schoolwork! They'd all gone out the other door.' But the nearest school was at Otamauri in one direction or Moawhango in the other 'and you couldn't get there in the winter anyway, so there was no alternative. I had Johnny on correspondence at five, and two littlies, Peter and Alan. And I still had to cook pretty often, with the cooks coming and going.' She would be given little notice about the extra mouths to feed. There were four or five shepherds and general hands in those days, and they would arrive up at the house for dinner whenever they came in and found the cook blotto and no meal ready for them. The boys went away to school just before their tenth birthday. 'It was awful when they all went away. But then in the school holidays I'd suddenly have to cook for five instead of two, and I'd have to treble the milk order and the bread order – but I didn't enjoy sending them to school.'

> It was very tough on Johnny because he was the first one. I think in his first term he lost a stone and a half. He wasn't eating – he just wasn't used to eating with a whole group of boys and couldn't cope with all those people around him. The matron looked like a dragon, but she had a heart of gold, and she looked after Johnny. She'd call us up and tell us what was going on. They took him off to the sick bay so he could have his meals in peace. He found it very hard because he didn't have much contact in those days: we only went to town every couple of months. They would stay with my mother sometimes when they went down, and visit Jack's mother at the flat in Hastings – and then they all moved out to Te Awanga and the boys went out there for exeats. But when Johnny first went to Hereworth there was only one one-day exeat a term, on a Sunday.¹⁰

Once they built a new cookhouse in the late 1960s they could have a 'much more reliable' married woman cook. 'A single male cook is out here

The ceremonial opening of Pohokura woolshed, 1977; Nanny Roberts 'cuts the ribbon', watched by grandsons Peter and Alan.

for all the wrong reasons: he's running from the law, or from his wife, or from the liquor, or from money problems. Cookie and Wi Matahuki were the first couple in the new cookhouse and they stayed four or five years. She was a good cook and he was a very good fencer – he put up miles and miles of fencing – 4½ miles in the Comet boundary fence alone.' Alan and Carolyn Clarke were in the cookhouse for a while, before they moved down to Pohokura when the road was improved to get timber out. Their house was moved in by truck from Hastings, and added to over the years. Alan, a general hand, fencer and tractor driver, was there about twelve years. Most of his family were born and raised down the valley. In July 1977 the woolshed that had been built down at Pohokura was ceremoniously opened by Winnie – Nanny Roberts as she was called by the family. The original cookhouse was still there and in use, and two prefabs were set up and used as shearers' quarters and for the scrub-cutters until new shearers' quarters were built.

Lance Kennett came to Timahanga after more than three years working at Ngamatea for Ray Birdsall and then Phil Mahoney, who took over from Ray, and then some time mustering in the South Island. Jack knew Lance and his father Alan, and offered Lance a job as head shepherd. After a couple of years he went on to a stock manager's job, and then back to

Jimmy's, 1959.

Timahanga as stock manager for Jack. He was married by this time, with two young children, and they lived at Jimmy's, halfway down the valley, where Jack had just put another house.

> I like it up there, that sort of country. Once you go there, work there, it always seems to draw you back for some reason. I spent about seven years up there between those two places. It was good experience, I enjoyed it, and I followed in the footsteps of my father. It's a property that has a lot to offer – there's hunting and fishing, and I learned to water-ski at Pohokura. Jack has a great set-up there on his lake and he'd always offer to take us down with them. It doesn't worry me being a bit isolated, but to live in a place like that you have to have the right type of lady because there's not a lot of social contact. My wife wasn't happy up there, 8 kilometres and eight gates to the road, and the long trip to town to see her folks, so we only stayed two years.
>
> Timahanga is more appealing to me than Ngamatea now. It's a bit slower paced, not large scale and mechanised like Ngamatea. It's different country, and Jack still uses horses, so the stock work's different. Jack has always retained his Corriedales and I think they're pretty happy with the results they've had out of them over the years. They were still cutting and burning scrub when I was there. The helicopter would drop something – a napalm and petrol mix I think – in droplets all through the scrub which would light it. But look at it today – a very good farm. All credit to Jack, he's done a great job of breaking it in. Nobody can take away what he's done at Timahanga.[11]

Bob and Bev Ralph worked at Timahanga for six or so years from the beginning of 1987. Bob had done agricultural work with crawler tractors, and was a very handy general hand, a good fencer and carpenter and a great fix-it man when it came to chainsaws and motor mowers. He started by repapering the cookhouse and their quarters while Bev became the station cook.

> I'd only ever cooked for the family, but cooking for the boys was no problem. There were usually two or three shepherds and one general, but up to a dozen at docking – I'd just do a tucker box for them for lunch and smokos – and sometimes there were casuals overnight if they were spraying or bridge building. Jack would come down for meals too when Jenny was away looking after her sick mother. He was very good to work for, very appreciative of everything we did – and he made sure the boys were good to me too. He'd always give me advance notice when there'd be extras for meals and he never checked up on Bob or interfered – just let him get on with whatever needed doing. And he gave me grazing for my sheep. I started adopting stray lambs and sometimes had up to eight at a time – Bob built me a feeder for them. We paid for the milk powder, and then Jack let me graze them. I got quite a little flock. Then I started spinning and dyeing wool and curing sheep skins, deer skins, possum skins. Timahanga was a wonderful place for doing things like that.[12]

The hot pool at Pohokura had reverted to its original state as a spring and needed digging out and relining if anyone was to have a relaxing soak in it, so Bob co-opted Matt Burke and Guy Matches, two young shepherds, to give him a hand with the job. Someone caught it on video and there they were with shovels and rocks and plastic lining creating their own private pool, and taking a dip when the job was done. When musterers arrived at Ngamatea in the early days and heard there was a hot pool down at Pohokura they thought, 'Oh yes, another Ngamatea story – but it was there all right. It didn't get too much use in our time – it was such an act to get there – but it had been walled in and you'd just clean it out and you had your own hot tub.'[13]

Johnny Roberts was back home working by the time the Ralphs left, but Alan was still at Lindisfarne. When Johnny married Jackie Wood, a classmate from Lincoln who had gone on to nursing, they moved into the house at Pohokura. Then Al married Karyn Hughes, who was also a nurse, and they moved to Jimmy's. The two families are a great combination. Even the two brothers are complementary: Johnny is inclined to tractors and machinery, Al to stock, horses and dogs. Jack and Jenny count themselves lucky to have two out of their three boys settled on the property.

Their mail arrives at the same mail-box as Ngamatea's, at the Otupae turn-off. 'We could go five times a week – to get the bills! We used to go twice, now we just go once on a Wednesday and it takes half a day sorting out the cookhouse mail, and the boys', and Johnny's and Alan's and half a dozen giveaway papers.' There is no point in them buying a daily paper – 'the news would be out of date by the time we got it' – so they rely on radio and TV for that, and if they need anything urgently they can get it brought out any weekday – and do the 34-kilometre round trip to the mail-box to pick it up.

The quality of life at Timahanga is influenced by the state of the road. They remember a time when the Annie was closed completely for three months while it was widened – 'they'd never get away with that today!' They could take their car so far, then carry Johnny and drag his pushchair over a great mound of rubble to where Jenny's father was waiting on the far side to drive them the rest of the way. With the Annie sealed and the Taruarau hill widened – 'and every time someone goes over the edge they widen it a bit more' – they can get out more easily, and visitors can get in. But so can everyone else.

> We're not just the first aid post, we're the information centre and the emergency phone, and the emergency petrol supply. We do have petrol here, but we don't fill them up. In winter we get a snowstorm, and then you get people pouring up from town to see the snow and they all think it's over the next hill, over the next hill. They get up here and of course they've got no petrol. That really annoys us. There's a notice at Fernhill saying how far it is to the next gas station, but they don't see that; it's surprising how few people see that sign. We had some people from England who had a campervan – a complete family travelling through this road. They had a look at the map and when they saw Timahanga Station, Ngamatea Station, they thought they were service stations and they could get petrol at each one. And there aren't many signs along this road after Fernhill so we have a lot of people call in who have no idea where they are, where the road goes to. They're trying to go to Wellington and they've missed the turn at Fernhill! And if it's night, or winter, it's a bit scary for some. We've had to put people up – quite a few. Luckily we've got the quarters. Some people have come in for petrol and we've given them some and told them to come in and camp in the quarters. Depends on the circumstances.[14]

Sometimes the police would phone telling them to look out for so-and-so coming through. 'One time they asked us to put a bulldozer in the middle of the road because they wanted to catch someone – and we did – and they did.' There is always a truck going off the road, or cars, especially in the ski season. The majority of the accidents 'are skiers or city people who have never driven on a metal road' and as the house is in sight of the road they end up on Jenny's doorstep wanting to use the phone or get patched up. 'There's a first aid post notice at the gate, and a St John first aid box on the wall outside anyone can get into if they need it and we're not here. I'm not a qualified first aider – I can do basic first aid, but that's it – but because I'm the only person around they asked me if they could put it there.'

One night, near the end of shearing, they had shedded up, shut down the generator and dropped into bed when they heard a rather noisy car drive up to the back door. They waited for a knock, but none came, and after a while Jack got up to investigate. He found a young fellow in his mid-twenties in a Volkswagen with the motor still running.

> He'd put a piece of plastic pipe on the exhaust and into the driver's side window and wound the windows up and he was almost unconscious there – suiciding on our doorstep – or trying to. I got the door open and shut off the motor and got the bloke out and rang the police. They said 'Keep him talking till we get there' – which was an hour and three-quarters later. Rob, our nephew, was here and he got up, and we didn't know if this bloke was anti-woman or what, so Jen was in the sitting room keeping out of sight, and we were filling him up with black coffee and talk, talk, talk, all the rubbish under the sun to try and keep him talking.[15]

It turned out that he had come to Timahanga to end it all because he once went hunting up in Boyd's bush, and it was the only time in his life that he had really enjoyed himself. 'The police took him down to Hastings. It was very sad, and we never heard the outcome of it all. Sometimes it's a cry for help, but we're not qualified to deal with that sort of thing. The first aid box doesn't cover that – nor does my first aid training, I might add. We never know what's going to happen next around here, and probably just as well.'[16]

The road down to Pohokura was not without its problems. When Jack freeholded the pastoral lease in 1964 the Comet blocks were cut off and went to Lands and Survey, and they surveyed out the road Lawrence had put in. A lot of people then regarded it as a public road and thought they had right of access.

> They just about put us off the place – poachers, people taking dogs through, horses. It went on for years. When Denis Marshall was Minister of Lands we had him and some other officials here and we took them down the road, and he said, 'We'll have to sort this out.' And he came up with the idea whereby we gave about 2,500 acres to the Crown – it was part of Te Koau and then

it was included in Ruahine Forest Park – and the road, all 60 acres of it, became Timahanga freehold. The public have the right to walk down the road provided they fulfil certain conditions; there are rules and regulations written out on a great notice board at the top end of the road. But first they have to ask our permission, and we can refuse it on the grounds that it is lambing or calving or high fire risk or stock movement on the road at shearing etc. They can only go on foot – no horses, bikes, motorbikes – so limited access subject to certain conditions. And of course OSH regulations come into it now too, and that's a real worry.[17]

The law has eased up a bit these days on you impounding vehicles trespassing on your own property, but you have to be careful. If you damage the vehicle you're liable. Before, we couldn't lock the gate to keep vehicles out – or in, if we caught them trespassing – but we can lock the gate now any time we like, and if a vehicle is on the wrong side of it, too bad. And our rights are marginally better protected vis-à-vis crooks and poachers and trespassers. At one time you could only prosecute for trespass if you caught them on the place twice in six months. Now, if they're carrying a firearm, or got a dog, or don't have permission, you can prosecute them for being illegally on your property. It's still a different story once you get them to court, though.

One night Jack saw lights going down to Pohokura, so set off in pursuit, accompanied by a former shepherd, and 'after quite a search', they found the vehicle.

> He'd driven it off the track that goes up into the bush, and he and his mate had actually cut scrub and stood it all round the vehicle as camouflage. I knew him – he was a sickness beneficiary, but that didn't stop him carting out deer! I think he was shooting for a helicopter. He was a bit of a nasty piece and he'd been troubling us for some time but all we could do was tell them to get the hell out of it. We couldn't take their rifles in those days. Anyway, he wasn't moving so we padlocked the road gate – and he cut his way out, and as he drove out he left a scratch of paint which they analysed and traced back to his car. When it got to court do you think for the life of me I could tell the judge that he cut his way *out*. He thought he was cutting his way *in*. We'd padlocked him in because we wanted the police to come out and catch him on the place. But they didn't come, wouldn't come – it's the middle of the night. They came eventually, in daylight, and got the paint. He was fined a couple of hundred dollars or something and he walked up to me outside the court and said, 'I'll be up in the next few days to shoot more deer to pay my fine.'
>
> The poaching got so bad here during the velvet season, round Labour weekend, that we never ever sent anybody out on a brown horse to muster a scrubby block, because it was just too dangerous. If they saw something

move in the scrub they'd shoot it. They reckon it takes 30 seconds to shoot a deer, get out of the chopper and pick up the carcass and away. Cattle could be targeted too. One of my shepherds saw a steer down and went to investigate and saw it had been shot and all the meat boned off it and the skin put back over it. If they don't get a deer they will sometimes take something else. We've had odd sheep shot and just the skin and the guts left in the paddock.[18]

Jack set up his own deer farm in 1978 by ring-fencing an area at the top of Timahanga, not too far from the house, where they could monitor the access. They actually fenced in Sikas and gradually worked them into two pens to one side of the deer farm, then increased the herd by pushing two pens into one and opening the outside gate. When wild Sikas came poking round the outside, then through the open gate to commune with their captive mates, Jack would slip up and close it on them.

> Sikas are more interesting than reds – fiery to handle. Most people don't farm them. We did it for the hell of it because they were so interesting. When you go up there the red deer tend to just drift off and walk into the scrub, but the Sikas are as cheeky as can be. Once they got to know the vehicle, after being fed hay from it, they'd actually come to within 3 or 4 metres while you were feeding out. They were like that out in the pens on the deer farm, but once you put them in the shed it was like trying to handle a bunch of firecrackers. When we built our deer shed the walls were two metres high and when we first started to handle the Sikas they were doing a standing jump and going from pen to pen over those walls – from a standing start, and with antlers on! We finished up having to put extra rails round all the pens in the shed.

They still farm reds but sold off the Sikas, and have reduced hind numbers in the past couple of years since TB, sporadic since 1991 in both cattle and deer, has resurfaced on the property. They were keeping the best of the hind fawns and a handful of stag fawns and selling the rest as weaners, but until the property is clear of TB again they cannot sell any stock. Overrun with possums, they mounted a big pest control programme in 1991, starting with a ring of bait stations with talon right round the deer farm.[19] A night or two later Jack and Jenny saw possums fifteen deep lining up to get to a bait station. 'After that they did a big 1080 drop through the bush and the birdlife now is fantastic – it just explodes, and all the young trees are regenerating. They've killed thousands and thousands of possums and now there's hardly a possum on the place, but we've still got TB. We think it's ferrets – and 1080 doesn't get all of them.' In the days when Ngamatea and Timahanga were plagued by rabbits, a government rabbiter living on the main road down by the Taruarau amused himself farming a dozen or

Jack and Jenny with their grandchildren and English friend Keith Brunskill at the lake at Pohokura.

so ferrets. When he had had enough of them, he liberated them. Whether or not today's ferrets are descendants of his cannot easily be proved, but he is a handy scapegoat.

The ex-musterers like to come back to Timahanga and reminisce about the old days. They even organise bus trips for various clubs they belong to and Jack and Jenny welcome them all, often down at the Pohokura woolshed and quarters where they can have a cup of tea and a history lesson from Jack. This tickles the old hands. Bill Wells remembered Jack as a boy: 'he was shy as anything, you wouldn't get boo out of him, and now he stood up there and gave a great talk about the station to all the old folks on the bus'. They remember him as his father's son: Bill Cummings thought him 'the dead spit of his dad – in his manner, everything', and according to Gordon Grant Jack is 'about as quiet as Lawrence was, but when he gets going he's really good to listen to'. Dixie McCarthy has the advantage of living quite close, in Taihape, and has always been able to keep in touch. 'I get on well with Jack and Jenny and always have a good yarn when I go out there. We've got Corriedales on our place just out of town here, and we buy our rams from Jack – about ten every year. But ever since Jack's been down at Timahanga I've done work for him. We've been contracting for about 60 years now; we did a lot of discing earlier on and now my son does a bit of work for him with the tractor.' Jack called on him in September 2003 when the endless rain brought down one huge slip and half a dozen smaller ones across the Pohokura road 'and did more damage than cyclone Bola'. Pohokura was cut off for three days, but with Dixie's tractor and Jack's they got the road open again, and 'now that Jack's got his own tractor, a little slip they can handle on their own'.

Peter van Dongen had such happy memories of Ngamatea that he always wanted to go back to the area, and after a stint dairy farming he and his wife Joy were able to buy 3,500-acre Kaingaroa station at Pukeokahu. 'I thought I'd come home, but I got very ill down there and Jack was a great comfort to me. He'd had open heart surgery and I had the same thing. He knew what I was going through, he was really good – just talking with me. I'll never forget that. And he was good for Joy too, and so was Jenny.' One day Jack and Jenny took the van Dongens for a drive through the property – the full tour, five hours and 75 kilometres, all on their farm roads.

Timahanga today is a different world from that which Jack inherited. He has been able to keep some of its historic buildings, but the landscape has changed beyond recognition. Nothing is easy up in the high country, and as the musterers say when they come back, 'what you see at Pohokura is mind-boggling when you compare it to the early days. The work that Jack and Jenny Roberts have done is outstanding, one of the land development success stories of New Zealand.'

Chapter Seven

The Development Years

You come here and you stay here and it's part of you forever.

When Ray Birdsall came back to Ngamatea as manager in October 1966 there was a stack of work to catch up. The place had gone back badly when Lawrence was sick: stock numbers were low, sheep were dying and there were only two men on the place – a tractor/truck driver, and a cowman; no stockman had stayed on over the winter. Jack and Margaret did not even know whether they would be able to stay farming.

Ray had been stock manager on Haupouri station out on the Hawke's Bay coast but he would call Lawrence now and then to ask if he could go shooting, usually in his Christmas break. He would leave his car at Timahanga for two or three days and go up onto Cameron Spur, along the top past the Log Cabin and over to the Rock. It was Margaret who called Ray and said they were looking for a manager for Ngamatea. Lawrence and Winnie had gone to Tauranga for a while, hoping the warmer weather would help, but now they were back in Hastings. Ray went to visit them and Lawrence told him the stock numbers had dropped, but neither of them appreciated how bad the situation was. Ray did not go up to Ngamatea before he agreed to take the job, although Margaret told him he would find things pretty grim. The trustees wanted to sell the Otupae block but Margaret asked them to hang fire: she had a good manager, and she was sure they would pull through. Ray's wife, Maxie, did not want to leave her beautiful frost-free coast for the frost and snow of the high country, 'but when the chance came for Ray to go back as manager, I knew I had to say yes'. They moved to Ngamatea on 1 October 1966 – and it was freezing, according to Maxie anyway; she wore all her clothes and a set of Ray's on top. And Anthony, who turned five on 5 October, came down with measles.

Ray was given complete control to run Ngamatea but there was simply no money for all the things he needed. He used his own chequebook to get a plough, hay-elevator, motorbike. He went out every weekend and shot deer in their hundreds and sold the skins and venison to buy saddles, welding equipment, TVs and even a truck for the station. Feral venison

Sun on the tussock; the Tit in the background.

was paying well, and the boys were encouraged to shoot in the weekends or when work stopped because of snow. They did well out of it and the station got 25 per cent of what they made from meat. 'Ray amazed me,' said Larry Cummings. 'He would always give us the pick of the blocks and take the last choice himself – and he'd always get the most deer!'[1] Ray would not have a general hand on the place: they cost too much and they did not pull their weight. He told Maxie she could not have a house dog: they did not earn their keep. Luckily she did not want one.

Margaret had promised them a new house if they took the job and Margaret and Terry, Jack and Jenny, Ray and Maxie had all been over to Rotorua to look at Lockwood houses. It took about three years for the replacement to materialise so meanwhile they came to terms with the idiosyncrasies of the old homestead which, at 3,200 feet, was the highest permanently occupied homestead in the North Island, possibly in the whole country. It had served its time.

> You'd hear this clop, clop, clop – possums on the roof, every night – and there was a hole in the bedroom ceiling and I looked up and there was this eye looking down at us lying in bed. And they would wet through the ceiling. I'd be cleaning up puddles around the place all the time. This day the stock agent came to the door and came in and just stood there. I said, 'I wouldn't stand there if I were you'. 'No, no, it's all right' – and he kept on talking. I insisted

he move a bit, and tried to explain, but he kept talking. And next thing down it came – he got piddled on. I had to clean him up.[2]

The new house was built just a little in front and to the right of the old one. It was 'just an ordinary three-bedroom house' but Maxie got her pantry and 'good big bins for flour and sugar' and there was the sleepout on the garage, and the cottage for summer use at least. 'It used to get extremely hot in summer and they had those beautiful big trees, right down to the ground, around the lawn at the homestead.' The best thing about the old house had been the Aga stove. It had had its day, but Ray converted another coke Aga to kero, and it was 'a fabulous, fabulous stove' – until he had to clean it and then there was soot from one end of the kitchen to the other. He got another Aga and put it in the cookhouse: one less job cutting cords and cords of firewood. When they pulled the old house down they found newspapers from the 1880s stuffed into door jambs and window sashes, 'with ads for farms on the east coast and oak tables imported from England at 1/- each!'

D'Arcy and Nat Fernandez came down for a while right at the start to give a hand with the books and the children's schooling, but true to Ray's do-it-yourself attitude he soon took over all the Ngamatea and Timahanga bookwork – the wages and the station stores. The children's schooling was a bit more of a problem. Tim had had three years at school and Anthony, not yet school age, had been allowed to go for one week for a taste of it before they left Haupouri. They were both on correspondence the next year and went through Standard 5; then each in turn went away to school for five years. Their first teacher was a nice young fellow from Canada, but he had only a one-year work permit, and despite the fact that it was so hard to get teachers on back-country stations they could not get his permit extended. Then there was a girl who favoured one boy over the other, and a man 'who touched the boys' legs', and finally 'a really good chap, a semi-retired school teacher from England. But he never stayed that long – hard to keep those sort of people.' Maxie took over at that stage, and she and Tim struggled with the new maths. 'The workbook was designed so the child could work on his own and the answers were in the back. Well, the first term the two of us together were getting 36 out of 100 and that was just terrible.' She was 'wailing about not being able to teach Tim and how he'd never be any good in life', and Ray just said, 'Right – we won't wait for next year – ring up Lindisfarne and see if you can get him in.' When Anthony came to Standard 6 and the terrible maths curriculum Maxie suggested she go down to the Correspondence School and learn how to do the maths. 'Oh no, Mum, I want to go to school. I've been so lonely.' With both boys away it was Maxie's turn to be lonely, but she almost never had the house to herself. There was the usual procession of stock agents and vets and wool classers and 'there were only six weeks in the winter when nobody came in. The snow went right across the fences and the icicles hung down around the house. It was like *Doctor Zhivago*'. Sometimes Margaret and Terry brought both their mothers up for the day, and once or twice Mrs Roberts and Mrs Apatu stayed over with Maxie, to their great delight.

Ray was 'a bit bewildered' by the Fernies' farming policies. He had done a lambing beat at Mangatapiri one year at Lawrence's request and 'their farming techniques were so bad all the hoggets were still running with the ewes while you were lambing'. But it was up to him now, and it was the beginning of the new mustering season so he just set about the old routine with what staff he had – two musterers from the previous season, Barry Dunn and Dave Andrews, and a few casuals until he could get a permanent team together. They had lost the grazing rights to the Harkness and the Manson, but there were a lot of sheep left on the Manson and one of the first jobs was to muster them across to the 40,000-acre Hogget block. The Hogget was Crown land, and someone would come out every couple of years to inspect it to see if it was suitable for settlement. 'What a lot of rubbish,' said Ray. 'They even sent me notices saying I had to keep the drains and spouting clean – and there were no buildings on it! It ran up to about 4,000 feet. Some years later they sent a chap up there to go out and look at it and he got lost out there and they had to go and look for him.'

The last full muster with packhorses was in 1968, then they straggled the back country for another two or three years and still took a packhorse or two to Golden Hills and Peter's. 'I had to go back as far as Kaimanawa Face and Gold Knob which goes down to Gold Creek, beyond Golden Hills. The Lake block was Ngamatea's, so was Peter's – but there were no fences. You couldn't fence that back boundary. There was a limit to how far the sheep went, though we used to keep finding them even out on Gold Knob.' They stopped growing oats in 1967. Jack and Ray reaped the last crop, stacked it in the ground pens in the woolshed and set up a chaff-cutter down there with a drive from the tractor or the shed, and a bit of ingenuity. From then on there was enough grass for horses on newly brought-in land or regrassed home paddocks. At the start they could only do a few of those a year and keep the rest for run-off because the newly sown country grew such terrific clover that there was always the danger of bloat. That passed once the pasture had been in a few years and ryegrass had taken over a bit from the clover. Ray always knew the place would grow good grass. Back when he was mustering he suddenly hopped off his horse one day, got out his knife and dug in the ground. The rest of the boys were rather surprised, especially when he said, 'Gee, this would grow grass, good grass.'[3]

A nephew of Ray's, Denis Goulding, mustered for him for two seasons from 1971. They were still grazing part of the back country – they took sheep out as far as Golden Hills, but not for much longer.

> We didn't muster the Hogget, the Manson, the Boyd, the Kaimanawas. We only straggled them, and we'd come in with about 150 sheep. Four or five of us went out on the main muster – Peter and Jack McLellan, Gavin Redford, Garry Coleman and me. We camped out there at Golden Hills, but instead of packhorses we used a little Ferguson 24 petrol tractor and Ray's car trailer – the cook took that and we rode out. There was a bulldozed track all the way out to Golden Hills that had been put in so a new Rabbit Board hut could be taken out there. The contractor who bulldozed the track would bulldoze all day, walk back to his truck at night, drive to his tractor, cook his tea and camp on the job – for a week or ten days. It's about 18 miles.[4]

The great challenge for Ray was to get the stock numbers up. There were about 12,000 sheep on the place, about 10,000 of them wethers out the back, some of them six-year-olds, and there were only about 1,000 ewes on Ngamatea. Jack sent up the English Leicesters from Pohokura and his next lot of cast-for-age ewes, and Ray put all of them, plus Lawrence's precious merinos, to fat-lamb sires. They needed every head of stock they could muster, and there was no way they could maintain the studs and go on breeding first-cross halfbreds. Ray set about stabilising the halfbred flock using their own halfbred rams. Margaret came up and she and Ray spent a whole day going through all the halfbred rams, culling them on wool quality. Ray went over them again, culled them on other traits and found the very best ram, which he put to the best ewes and started a new breeding programme – his own little halfbred stud based on that top ram. The next year the first ever fat lambs went off Ngamatea – Ray's lot, plus all the wether lambs from Timahanga which had been sent up for fattening, and those from Mounganui, another Fernie property near Moawhango.

In two years Ray had made a profit; after seven he was surveying his handiwork one day and said to Maxie, 'Anyone could come and take over now and it will never go back'. Ray had started a grassing and cropping programme with the only tractors they had – an old D2 Caterpillar with five-a-side giant discs and then an HD6 they got from Jack, which was a help in the 350-acre paddock they had started in. Ray had called up another nephew, John Bennett, to give him a hand with the tractor work. 'I had a marvellous time, the best eighteen months I've spent in my life. I drove the truck too – a J6 Bedford four-wheel drive, and helped with dipping, a bit of stock work.'[5] In the second year they got a new tractor, a small Ford County. Ray upgraded that with a bigger motor and they got a Same Drago as well, then kept upgrading over the years as the cash flow improved. As soon as they could they bought a 150 Ford County, which went down to Mounganui with the Same Drago so they could put in crop and fatten their own lambs, and Ngamatea got a 225 John Deere pivot steer, a good tractor in snow and well suited to development of tussock country because of the high clearance.

Ray began by developing a couple of blocks to the west of the station road, running down towards White's; they called them Stewart's, for the contractor who did the tractor work. 'It was good strong tussock land and there was good grass in White's, so we turned over the tussock and put it straight into grass – a 600-acre block, about 400 of which they could plough.' The next year they did the southern end of White's, closest to the part of Otupae between them and the main road – 'the best land, all shell rock, and tall snowgrass as high as your horse'. After that they moved east and developed Owhaoko D5 No. 2, towards the station road, and beyond, to the top of the Gully block. This was part of Owhaoko D5 No. 4, a 5,500-acre block in which Terry Apatu, through his father, had been a major shareholder. He was able to buy the remaining shares and freehold the block and make it part of the station.[6] Each year they were turning about 1,000 acres of tussock into grass and 1,000 acres of tussock or old grass into crop. They kept the tractor driver going year round, except for two or three winter months when the ground was too frozen. Some of the blocks were so big he could cut only one round a day for about the first week.

Rabbits were still a problem and three Hawke's Bay Rabbit Board rabbiters were based on the station. They worked across all the back country as well as Timahanga and Otupae, and camped at Hawkins' hut out towards the Rangitikei, or down in White's Valley. Life was pretty basic in a Rabbit Board camp but the huts were more substantial than musterers' huts and two of them were new. Larry Cummings, son of an earlier head musterer, Bill Cummings, started as a rabbiter in October 1969. The team spent a month out the back, a month in the shearers' quarters at the station, then moved down to the new hut at White's. 'We were chasing rabbits in the tussock and a lot of the scrub faces and old scrub burns – a haven for rabbits in the grassland itself at that stage. I think two of us got about 500 rabbits off one small area in a month.' The biggest job though was feeding the dogs, 40 of them – and remembering their names. The station supplied the dog tucker, mostly old rams, and they could shoot a deer a week. At Christmas their boss sent the two young rabbiters home with half the dogs each and Larry spent his break killing his father's old rams to feed them. Enough was enough. He told his boss to come and pick up the dogs: he was not coming back. Early in the New Year a job came up back on the station and Ray, who knew both Larry and Bill, gave Larry the chance to follow in his father's footsteps, although working mainly with tractors, not stock.

> I was still only a lad out of school, but Ray was excellent to work for. It wasn't a full-time job on the tractor. We put in about 500 acres of grass, and there was a bit of feeding out, and I just did whatever else was necessary – fencing, firewood, dog tucker, dagging. There were only six of us – Ray, three shepherds, a cook and me. But there was something about Ngamatea. Young shepherds wanted at least a season on there. I teamed up with one of them, Les Hazeldine, who I think was there just for the hunting. There was a good head shepherd; he wasn't into hunting, he was out training heading dogs. That was his spare time job.[7]

By 1972, when Ngamatea and Timahanga were split into two separate stations, Ray's stock numbers were well up and increasing by about 5,000 stock units per year. He had 350 heifers raised from calves and he had culled only three of them; the steers they raised were income. When Walter Fernie died in 1975 Terry and Ray went down to Otamoa and brought all the lambs back to Ngamatea for fattening, but the only stock they bought in were those first fat-lamb sires plus bulls every year, and later three Booroola rams – merinos carrying the Booroola fecundity gene, bred on the southern tablelands of New South Wales. Wool was the main income as it always had been, and increasingly lambs, but lamb prices were not high. They sold to Ambles, big Auckland dealers, who came and picked the lambs in the first year, but after that left it to Ray who had as good an eye as they did. A truck and trailer went out Wednesdays and Saturdays from January to May as long as there were lambs to ship – and until the works were offering a better deal.

John Brenton-Rule was shepherding for Ray from 1969 to 1971 and Denis Gayton and Peter McLellan were there for part of that time. Two of them would stay through the winter, feeding out and doing whatever was needed, such as giving Ray a hand to put in a new shower dip that took cattle as well as sheep. It was just up behind the covered yards but they still used the old swim dip for a few years; with big sheep and a large enough team it was faster than the spray dip. Roads and dams were put in as the development went along, and they built an airstrip and super bin up the top, by the station road, and flew the super on. On Stewart's and other early blocks there were tussock gullies that grew great grass when it was supered from the air, but when they changed to ground spreading with trucks the super did not reach those gullies and in a few years they had reverted. The plane was fine when they were doing long runs right down over the station, but when they were working across the face of the slope, with lots of turns back and forth, it got too expensive. There were advantages and disadvantages with both methods, but ground spreading was cheaper, and most of the country was driveable anyway.

One year Ray made silage but they lacked the gear for efficient feeding out, so went back to hay made by John Woollaston, a Taihape contractor like Peter Law, who had done the Ngamatea hay in the early 1960s. He had two New Holland hay bailers; he operated one and Noel Merwood the other. The first year Noel was there the hay was cut at the end of January, early February and was nearly ready to bail when a cold front came through and it snowed – a seasonal hazard on that high country. The hay was buried under 5 or 6 inches of snow and they could not work it for two or three days. Then he had to ted it again several times before it could be bailed and all that reduced the value by 70–80 per cent.[8] There were no general hands at Ngamatea to cart and stack the hay in Ray's time: 'my shepherds became all-round handymen – electricians, plumbers, concreters. They were there all year round, and just did everything – four shepherds, sometimes only two or three.'

Rex Tobeck and Kevin Schimanski, early mobile farriers, had taken over from Alan Kennett in the early 1960s; Alan had shod the horses in the old blacksmith's shop at Ngamatea for some years after Forbie Minto's reign. They lived in Wanganui but Rex worked out of Taihape all week and came out to the station every six weeks for about ten years from 1965. He and his apprentice Wayne Parsons or sometimes Kevin would arrive early at Ngamatea, have breakfast at the cookhouse and spend the morning shoeing 15 or 20 horses. After lunch they would go down to Timahanga and Pohokura for the afternoon then back to Otupae for the night. They were featured on a programme called *Looking at New Zealand* which was on TV in the days before *Country Calendar*. 'It showed quite a lot of Ngamatea, and the beautiful stables.'[9]

There was a fencer on the station almost year-round. Ash Dunstan, who had come to Ngamatea as a lad in the early 1940s, started just days after Ray took over in October 1966 and was still around, on and off, three years later. He had been fencing for many years. Among other jobs he put in a new telephone line for the station, 'a proper line with two wires'.

> I remember saying to Ash at tea one night 'How about putting my telephone poles in, Ash?' 'Yeah, right oh – how deep do they go?' We went outside after the meal. I said, 'Here's the pole – I've put the cross-arm on, the cups are on. I'll put a little square on the ground where I want the pole. How much a pole?' He grabbed his spade, ripped a hole in the ground, grabbed a pole threw it in. 'A dollar a pole,' I think he said – it took him so many minutes to do one pole. These fencers are pretty good. He wasn't using a posthole digger or anything

– good digging there, of course. So Ash did this big line across the farm, following the old Taihape road and over the top out to the main road.[10]

Eric Brooking came in 1975 to do 4 miles of fencing along the road, both sides of the Ngamatea gate. He stayed fifteen years and did close to 200 miles – 'plus all the short fences I did – I never counted those. That was on an hourly rate – not worth putting short fences on a contract. I worked up to 8 miles out from the station, fencing land being brought in from tussock and I used to fence right around and cut it all into paddocks. Some laneways – double fencing.' He worked alone, from dawn to dusk, seven days a week, by choice, with his trusty old Land Rover but no other means of communication back to the station. 'The first nine years I worked every day, but each month I'd take a long weekend, four days, for shopping down here in the Bay. No supermarkets Sunday then. After that I knocked it back to six days a week.' He had a posthole borer, but preferred to use a spade – 'that pumice country is soft, you can dig it – best ground I ever fenced on'.[11]

Fencers did not usually stay over the winter, but Eric did: 'As long as the snow wasn't too heavy I'd just fence on'. He built five sheep yards and a couple of cattle yards, one in Top White's in October 1980 and one on the Otupae block on the south side of the Napier–Taihape road, and when the weather really was bad he could work inside making gates for the yards. His wife Hazel was with him and they stayed in the shearers' quarters for a start, then the old cookhouse until early 1980, when a new Lockwood house was built for them, up on a bit of a rise, looking over the station buildings and way out to the back country. 'We had the opening of the house when we first went in and I put down a hāngi for the whole station. It was snowing that day. Hard job lighting the fire. In the end we got it going – turned out a beautiful hāngi.' Eric and Hazel, both keen gardeners, grew all their vegetables in a netted garden to keep the rabbits and possums at bay. Eric left his mark on the station, and three paddocks were named for him: Eric's, Top Brooking's and Bottom Brooking's. That really pleased him.

New shepherds' quarters were built in about 1971 – a six-roomed Lockwood – and in 1975 they built a new Lockwood cookhouse and from then on had married cooks. Jack Aldridge had stayed right through from July 1961 to March 1967 – a long stint for a station cook. He had been a 'good one too, very clean and fastidious, and he always made a special effort when the musterers returned from the back country'.[12] The first married couple in the new cookhouse were Anne and Chris Forrester; they were followed by Chris and Des Page in December 1976. The Pages were both townies but had had a year dairying, which did not suit them, and decided they would be cook and tractor driver, although they had no real experience of either job. But everything about Ngamatea suited them. Chris normally cooked for 4 or 5, and up to 20 or so at busy times, and 'just loved it, although it took a little while to get used to the appetites, getting the quantities right'. The Brookings were still in the old cookhouse, and Hazel would drop in to see how Chris was getting along. 'She was lovely. She used to make us Maori bread now and again for a treat, and if I was peeling potatoes or bottling or whatever when she came over she would just pick up a peeler.' Chris bottled cases of fruit, but a lot of this work was done in the evening. 'The boys would have their dinner and then we'd all congregate round the kitchen table and they would just peel and laugh and talk. It always amazed me – after a day's work. I would never have asked them. But they knew I was doing it for them so they all pitched in.'[13]

Chris had her first baby just about a year after they went there – right at shearing time. She worked until two or three weeks before the baby was due, then went down to Hastings for the birth. But she was back within a week, and cooking again right away – on the coal range; the power was off for about three days. Fortunately Rebecca, the first baby at Ngamatea in almost 40 years, since Margaret was born, was a good child. Maxie and Hazel took turns looking after her whenever they got the chance. 'They both loved the baby, because really it was a man's world up there, and to have a baby was something special.' When Rebecca was just a few weeks old, the district nurse 'drove all the way out from Napier just to see us. Only the once. I was coping really well and Rebecca was doing fine, so she didn't come back. I used to take her for a check when I was down in town. We got one long weekend a month to have a town day.' They seldom stayed over when they went to town; they could not get back to Ngamatea fast enough. Des thought 'it was just a world of its own up there – like the outside world wasn't important. Once you were living there what happened outside the fence didn't matter. It got to the stage that even at Christmas the guys wouldn't go away, Chris put on such a good Christmas for them.'

Des became offsider to Dan McDermott, the tractor driver, but the long hours and days did not suit him and after six or eight months he was able to change to shepherding when Bruce Fisher left. He offered his dogs to Des, who shepherded then for the next 20 years. Ali Robertson was head shepherd, then Kerry Henaghan took over; Kerry was on Ngamatea over nine years. Craig Hammond was there too – another second-generation stockman on the station. Lance Kennett was also there. He was not quite seventeen when he persuaded his parents to take him up to see Ngamatea and ask Ray about a job. One came up after four or five months, just in time. It was almost the end of an era; Ray left about eight months later when he drew a farm in the last of the Lands and Survey ballots. The stock work was still done on horses but Ray had put in a lot of four-wheel drive tracks and at times the boys would be dropped out the back with their dogs and walk back in, so Lance got to muster some of the closer-in country his father had mustered, and with Kerry Henaghan he straggle mustered further out.

❦

Throughout the 1960s the Crown continued to eye Ngamatea covetously. In August 1967 it had noted the successful development on Otupae and Mangaohane, and at the end of the year was thinking about development of 'the large tract of unalienated Maori land over which Owhaoko station have free grazing', as though the development of lower, accessible stations in the district meant the same could happen on Ngamatea's high and remote back country. They were enlightened a few months later by yet another report on the station:

> The country varies from good tussock (silver and snow) to poor tussock, scab weed, mosses and lichens, an area of swamp and poor pumice flats, to steep badly eroded mountainous country, much of which is still covered in small mountain beech. Soils are pumice while the gullies and flats have up to 20ft of pure pumice over rock. The steeper hills have had all the pumice washed from them exposing greywacke rock and sandy soils on the gentler slopes – mainly Kaimanawa and Urewera sandy silt of low fertility and subject to erosion, with only fair phosphate response.
>
> All the country suggested for development is over 3000ft and in winter is completely frozen over for days at a time, and snow can lie for up to 3 weeks or more. Spring growth comes away at the end of September and ceases again at the end of May. For the remainder of the year stock have to be fed crop or hay or allowed out on tussock country to forage. Rainfall is only about 30" per annum.
>
> The land being grazed by Owhaoko Station carries about 700 sheep on approximately 16,500 acres over the winter months. The station has no legal tenure, but has its use for the payment of Rabbit Board rates.

The Crown was not about to give up, though. Nearly two years later it thought perhaps 'the Swamp could be drained', although 'the most suitable land for development' was Owhaoko D5 No. 3 and D5 No. 4, leased by Ngamatea and both fronting the Napier–Taihape road. Ascertaining the cost of development would be a lengthy and difficult affair: 'it took an officer two weeks just to look over the land many years ago'. Aerial photos would be required, Maori Affairs, Lands and Survey and Forest Service officers would have to go into the economics – and 'a base farm' would still be needed. The Crown had Owhaoko D5 No. 4 in mind: the lease was due to expire in 1972 so it was suggested that any further action be deferred for a couple of years. No more was heard about this; by then Ngamatea had freeholded the block, and it was under development. But the Crown did note that, since investigations had begun, the 8,897-acre Owhaoko C3b block – part of which was Kaimoko, or Boyd's bush – had been 'sold by the [Maori] owners to one of their number – a relative by marriage to the lessee of Owhaoko D5 No. 4'.[14] It had taken a while for any of that bush to become part of Ngamatea. A big hui had been held at Omahu on 6 November 1934 to consider a proposal from Drummond Fernie to buy part of the bush on the adjacent Owhaoko C7, 'but the meeting ended abruptly due to confusion among the respective beneficiaries ... and Mr Fernie not having the necessary papers to clear the position'.[15]

At the end of 1969 a Lands and Survey officer was complaining that 'Apatu and Roberts who now own Boyd's bush have let milling rights over part to Robert Holt and Sons, Napier. We could do nothing to stop this.'[16] The milling was short term, designed to pay for a properly formed and aligned metal road in to Ngamatea and a bit of building at Timahanga, and when Jack took over the block in 1972 he vested it in the Queen Elizabeth II Trust. That did not get a mention in the records.

The Crown was concerned that if 'the key blocks within the complex' were freeholded this would 'effectively restrict any development of the land'. It wanted the titles researched and the Crown's interest registered against them 'to restrict the further whittling away of the best and key areas' – although the Crown had never done any sort of development up there, and Ngamatea was developing all possible land. This must eventually have come to the Crown's notice: in a few years it was pleased to arrange with Margaret the exchange of a small area of Crown lease, truly a key block that could be developed, for a much larger mountainous and gorgy block that could never be developed. The Crown block was D6 No. 2, right in the middle of the property between the Homestead block and the Swamp; the land the Crown took was on the north-west boundary above the Rangitikei and it became part of Kaimanawa Forest Park. Sanity was starting to prevail.

All that remained was for the Crown to return to the Maori owners the land that had been gifted but never used for the settlement of returned Maori servicemen. In May 1969 the Crown wanted a good part of the A blocks on the upper Ngaruroro transferred to the Forest Service for conservation purposes; they had already taken 6,833 acres of the northern part of A East for state forest in 1939. In 1956 some of the Tuwharetoa owners were asking that the rest of the land be returned to them. By 1971 the government was considering special legislation 'if necessary' for the return of the land. Two years later the Forest Service, particularly concerned with soil conservation and river protection, hoped Tuwharetoa might agree to the land remaining in Crown hands. But 'the owners fairly firmly agree they want the land back. They may then be prepared to talk business with the Crown. Owners' opinion must prevail.'[17] A few weeks later the Minister of Maori Affairs, Matiu Rata, announced at Otukou Pa, Rotoaira, that the government was to return a 16,000-acre block – part of A East – to the owners.[18] The process continued until the rest of the Gift Block was returned in October 1974, but the Crown retained parts of the land bordering the Mangamaire and Ngaruroro rivers.

In 1951 when Ray Birdsall first went to Ngamatea there was no power. He remembered shedding up after 8 p.m. with 'a little lantern' in his hand, banging his head as they worked in the woolshed, and using candles in the quarters until the station's own hydro dam was finished. It was built on the Woolwash Stream, the outlet for the great Ngamatea Swamp, where the remains of an old woolwash dating from the Studholmes' time could still be seen. Noel Roberts was around then with his bulldozer and took a lot of big old beams out of the Woolwash to prevent them fouling up the intake for the power scheme. An engineer from Raetihi, Colin Gillett, inevitably known as Razor Blades, got the building contract.

> There were that many men there – casual labourers for the station cutting the oats and chaff, all the electricians, a couple of Aussie chaps working under the woolshed making the bricks from pumice to build the power house. Both tables in the old cookhouse were full and after tea we used to go out and play a game of rugby, two teams. They built the big dam first and they made a good job of it, then they'd join the big steel pipes together and concrete them in place where there were outcrops of rock. This Gillett would come racing across the top of the dam in his tennis shoes with a wheelbarrow full of concrete and he'd come racing down the pipe – this is true – and the seams of the pipe were three or four inches high where the bolts went through right round it and he'd jump over these things. One day he came racing down, and over he went, down onto the rocks, the barrow of concrete on top of him! He had an apprentice electrician, Colin Cribb, doing all the wiring with the others. He was a good joker; I made friends with him.[19]

Flood on the Woolwash.

They built a power house down in the gully below the dam with a 35- or 36-kilowatt generator and 'a few horseshoes on the governor to balance it; this Gillett was a real rough joker'. The scheme worked well enough for some few years until a cloudburst dumped inches of rain on the station and the flood came right over the powerhouse and ruined the generator. Another contractor, Norm Andrews, built a new power house only halfway down the hill with a 65-kilowatt generator. It was never totally satisfactory; in winter there was enough power to heat and light the whole of Ngamatea (it never was connected to Timahanga); in summer there would be a water shortage and they could not always run it. In Alan Bond's time, in the early 1960s, 'it used to play up and you'd be up to your waist in water trying to slow the turbine down. It used to get some rate of knots on, specially in a flood. You had to get down into the house where the turbine was and get this great big beam and stick it underneath the flywheel and try to slow it down. That was one job I never looked forward to.'

When Ray came back as manager it was costing the station a fortune to bring an electrician out all the time to keep the power going, so he got his mate Colin Cribb to come out and teach him how to do the job himself. 'He had his own business by then and he lent me a lot of gear and taught me how electricity was generated – direct current to alternating current and polarity and all that, and from then on I ran the power scheme and the only help I needed was to renew bearings.' There was still a lot of

The powerhouse flooded.

work to running it. The odd eel would get through the sieve on the intake and Ray would have to take the turbine to bits; the big generator used so much water he would often have to go down and switch it off at night in the summer to make sure there would be enough water to generate power all the next day. At shearing, when use peaked, he would put it off at nine or ten at night and be back down there at four in the morning so the cook could make the early cup of tea. 'This went on for years. Then we put in a Lister diesel startamatic, about 14 kilowatts, and all I had to do was throw a switch in a little shed by the woolshed and it would start when the first one put a switch on in the morning.'

Then the wretched generator caught fire and all the wiring burned out. Ray was sitting watching TV on a Sunday afternoon when the set slowly died and all the lights went out.

> I jumped in the old Land Rover and tore over to see what was wrong. Got to the top of the hill and I could see the smoke coming up out of the power house. We had to get the crawler tractor down there; it's a steep hill. We made up a sledge, jacked the generator up and got the sledge underneath and hauled it out. It had to go to Palmerston North and they rewired it – it didn't cost that much. John Bennett was the tractor driver in those days and he was driving the truck, and brought it back and we put it in and everything went again as good as gold. We never even got an electrician out! I don't know what OSH

and all those would say these days if you did these things – fair go. But you had to be self-sufficient in those days. That was the trouble before – they weren't self-sufficient and you had to pay all that travelling for people to come out from Taihape. And the frosts – every year at shearing time I'd be in the shearers' quarters fixing all the burst pipes.

The mains power was put in in 1972. 'It cost us $70,000 something for the privilege of having the power.' The next year it got to Timahanga and a decade after that it got down to Pohokura, but in the interval Jack took the startamatic down there and they used it with the diesel generator that powered the woolshed and cookhouse.

By 1978 Ray had worked on Ngamatea as musterer, head musterer and manager for just on 20 years and was beginning to wonder how long he should continue. 'My life was at Ngamatea, but I was getting on – close to 50. These big places a long way from town – I think you want to be young-ish. As you get older it tells on you.' He decided to ballot for a farm, missed out the first year and said he would never do it again, but the next year a farm overlooking the western bays at Taupo came up for ballot. Maxie liked the idea: 'I have to have a view,' she said. 'You can't farm a view' said Ray. He went up to have a look at the place and was ambivalent, but then drew it in the ballot in December 1978. They gave him a week or two to make up his mind, and the deciding factor was a call from a neighbour: 'Don't turn it down, Ray. You get up in the morning and you look out the window and you say "That bit of grass is mine".' That was just what Ray needed, but son Anthony still was not enthusiastic. He had just left school and started working at Ngamatea, and loved the place. Margaret and Terry did not want Ray to leave either and even suggested he stay and just supervise the station without doing the physical work on it but they could not persuade him. He stayed another two months, until a new manager was appointed and ready to take over, and they left on 28 February 1979. The staff put on a farewell party at the shearers' quarters and invited everyone from the neighbouring stations. Hazel Brooking and Chris Page came up to give Maxie a hand with the last clean-up, the boys helped to move all the furniture and Des Page drove a load of bits and pieces up to the new farm. 'Margaret and Terry were very good to me,' said Ray. 'We parted really good friends and they gave me enough stuff to run my farm for a year – shearing gear, docking gear, all that sort of stuff.' Maxie cried, but on the quiet so no one would know. 'Leaving Ngamatea was traumatic – for both of us. Okay, we had a place of our own, but I just love Ngamatea. We still watch the weather – we can see right down the Kaimanawas – and we say Ngamatea will be getting this lot – they'll be getting snow.'

Margaret and Terry had come back from the north in 1966 and settled in Waipukurau where Terry was a farm adviser. 'I had a good job and a good salary, but we didn't live the lifestyle – it was a while before we built the Lakeview Road house. A Dutchman drew up the plans to Margaret's specifications. She'd had a collection of house designs and plans for years.' They had three children: Kate was born up north, and Renata and Nathan three years and six years later in Waipukurau. They got up to Ngamatea when they could, but only on day trips. After Walter Fernie's death in February 1975 Terry dispensed with the old accountants, appointed new ones and involved himself more with the financial side of station business.[20] It was the start of the era of government incentives – land development loans, livestock retention schemes, supplementary minimum payments, tax breaks and so on – the more land developed and the more stock carried, the bigger the payments. In the 1980s the incentives were in full swing and all the stations in the area were in development mode.

Ray's successor, Phil Mahoney, was appointed the new manager and was at Ngamatea until December 1990. He had grown up in Feilding and had the idea he would like to be a stock agent, but the firm he approached suggested he have a year on a farm, then come back and talk about it again. But he kept farming, married Elaine Shannon, who had grown up in the backblocks towards Rangiwahia and been on farms all her life, and eventually they ended up on the Kelly block, then an outstation of Mangaohane. A couple of years later Mangaohane was sold; Jim Bull bought the front block and Warren Plimmer the Kelly block, and the Mahoneys stayed on for another five years. Elaine was capable and energetic: she could farm, cook, sew, garden and do almost anything else. They had five children and when they went to Ngamatea they were overflowing Maxie's 'ordinary three-bedroom house' so another storey was put on to accommodate them all. There was a permanent staff of four when they arrived, twelve when they left – 'and that didn't include the builders and all the rest who came and went'. The six-bedroom single-men's quarters were soon too small and in 1982 a new eight-room Lockwood was built for the permanent staff. They filled the cookhouse and the overflow ate at the homestead. 'It would be fair to say in our entire time there very rarely did we ever sit down to our kitchen table to a meal on our own, as a family. It would be nothing to have eight or nine extras.'[21]

Seven years on the Kelly block working under Peter Green had given the Mahoneys some insight into the high country, but Phil would not have Ray coming back every second month to keep an eye on things as Terry wanted. 'He even hated me talking to the staff. He had good staff and they'd been

THE DEVELOPMENT YEARS

there for a while. Kerry Henaghan, Marty Crafar, John Roberts – they were all good stockmen, but he frustrated them. He was the manager, though, and if things went wrong his head was on the block.'[22] But with finance readily available, continuing development on Ngamatea would not be a problem. Initially Margaret and Terry came up frequently, sometimes bringing Winnie with them. Elaine remembered the first time she met her: 'I was so nervous, thinking, Gosh, this woman is an icon'. Winnie would have been amazed by such a reaction.

Phil, wanting to make his mark on the place, began with a big tidy-up. On 17 May 1979 he put a match to the stables. Three weeks later the great shelter belts around the house were felled and all the old outhouses – homestead stables, dairy, henhouse, cowbail – were bulldozed and the old orchard felled and burnt. It was all part of 'the great homestead clean-up'.

> In March we began to clean out the old stables and the implement shed, which unfortunately was too sad a sight to be able to repair. The stables weren't quite so bad, but borer had taken hold and there was really not enough you could do. It was far too big a job to try to restore it – it was a rebuild job. Yes, it was a historic building, but then like a lot of things the timber at that stage was not treated. It was a huge job. We had two very large piles – things that weren't salvageable, and things that were, out of the stables and the workshops and the blacksmith's shop. There were a number of things there that were salvageable, and down behind the quarters along the big macrocarpa trees, there were the oat thrashers and the old standing bailer. All that material, anything that was salvageable, we gave to the Taihape Museum. It all went in there, because it was part of the local history. We had talked with Margaret about it. She wondered if it should go to Hawke's Bay, but we said, Gosh no, there were too many Taihape people involved. We tidied up that old machinery and I think they carted away about three truck and trailer loads of varying bits. The rest went into a very large hole that was dug at the end.[23]

The new workshop combined with implement-shed, chiller-room and stables, which the Mahoneys designed themselves, were started in July and finished in November. In the middle of 1980 they started demolition of the old cookhouse. Built in the 1930s, it was practically the last of Tom Whimp's handiwork, and its old brick bread oven was still intact. The original shearers' quarters remained.

The Mahoneys then turned their attention to the sheep. They bought the Kelly block five-year-old Romney ewes and for a couple of years crossed them with the Booroola, then used the best rams from that cross with the Ngamatea flock. Wool quality was sacrificed in order to 'breed a sheep that suited the conditions. The merinos and Corriedales of the past were fine for the back country, but as it developed the grass got too good for them and they didn't thrive as well.'[24] Sheep numbers were growing, and at shearing in January 1982 a historic milestone was reached: Ngamatea branded bale number 1,000; in January 1984 they branded 1,227 bales from 62,500 sheep shorn.

In September 1979 Chris and Des Page left. There had been too many changes too quickly and they felt Ngamatea was losing its character.

> Before, you didn't care what went on outside the gate and with Phil coming in it was like the outside was coming in and starting to take a hand in the place. ... We were there when he burned the stables. They were looking at building a big new workshop for the tractors, which took priority over the horses. No new stable was built then, and we were still using horses. They were never stabled though – they lived outside, no covers or anything. They were a tough breed. We used the old stables – we'd saddle up in there in the morning. When the stables went we tied the horse up to the fence and saddled up there. There was a lot of history in those stables. To me it was devastating to see them go just like that. I think all the guys there felt the same way. It's not like they weren't usable. They weren't decrepit. The implement shed was getting pretty creaky but the stables were still as sound as. It just seemed a shame to take something with that sort of history. Didn't even take them apart – all gone in an hour – just ended up a pile of burnt corrugated iron.[25]

In March 1981 nineteen-year-old Richard Whittington became tractor driver on the station. He had never been up that way, never heard of Ngamatea, but a mate told him the job was advertised so he phoned Phil and went up for an interview. He was told that once you crossed the Taruarau you were on Ngamatea. 'And it's all this scrub country: climbing up out of the Taruarau you get up on top and it all opens out to a plateau. Name at the gate, drove in – and my eyes fell out. I hadn't been on a farm that big before. But what I was going up for was the tractors. I'd been driving John Deeres for several years – since before I left school – but this was by far the biggest one I'd had the chance to drive, and it was virtually new.' Richard was also a qualified mechanic so he became offsider to Murray McPherson, the station mechanic. In January 1983 Richard moved into the new single-men's quarters; he stayed on Ngamatea for almost three years.

> They were developing the block below the Tit ridge and my first day on the job Phil Mahoney came out for a round with me to open up the block and show me what he wanted done, because there were no fences or anything, no

streams or gullies as the boundary. It took us about four and a half hours to do about two-thirds of the round, as high and as far back as we could. Early afternoon he got out and said, 'What are you going to do?' – and I laughed and said, 'I'll do another round'.

So that was my first day. It was a real eye-opener, but I was happy to be in such a big tractor – 8440 John Deere, pivot steer, 215 horse power. That was a big tractor in its day. I probably went to the end of May developing. After the first year we got into it properly, doing 1,000 odd acres a year. You'd super-giant disc 1,000 acres, then off-set disc it once,[26] and leave it over the winter half broken down and you'd go back to the 1,000 acres that had been fallow for a year and work that right up – give it a couple of cuts with off-sets and heavy harrows behind. After a period of time, especially with snow and rain, the tussock just breaks down.

We just had 12-foot heavy harrows at the start, two sets together so 24 feet of harrows, to try and level it. Then we got a rotatiller.[27] Once we had that you'd have three rollers, 11 or 12 feet wide, and you'd tow two behind, one to each side, so you'd be doing about 35 feet at a time, sowing grass – three seed-boxes. Southern Haulage used to put the fertiliser on in bulk with trucks – 6 hundredweight to the acre, really poured it on, but it needed it, there was virtually no topsoil – then sow it down in grass, and being virgin ground it just grew and grew. The clover was up over the bumper of the Land Cruiser. When we got the paddocks smoother we'd take a cut of hay off them.[28]

The sowing needed to be done before Christmas and that meant long hours. In one week, his longest, Richard did 100 hours. He would cultivate during the day and sow from about nine or ten at night until two or three o'clock in the morning while there was no wind. 'Quite often before Christmas there'd be the old north-west winds and you couldn't really sow when it was blowing because the grass seed would just get blown off and you'd get gaps, and of course you'd get a hard time about that.' Around Labour Weekend he would work up a couple of paddocks of old grass and put in 80 or so acres of swede-turnips for the winter and the following year that would go back into grass. 'But you'd do that in a matter of hours.'

Phil did not worry what hours Richard put in, as long as the job was done. 'I never had any problem with Phil. He knew I was getting on with the job and if things weren't breaking down and there wasn't a problem I could go ages with no more than a morning hello to him.' Phil was an excellent organiser who could run the shearing, the contractors and the fencers all at once – a very necessary attribute at that stage of the development of the property – but not everyone got on well with him. He was no stockman and his approach did not always make for happy personal relationships. Paul Hughes, who was a full-time vet in Taihape from 1979, got on well with Phil and spent time with him and Terry, doing more than vets usually got to do on big stations. Phil, a keen deer man, would take Paul to stag sales and also to see how things were going with the cattle grazing out in the tussock.[29] For Paul it was the beginning of a long and continuing association with Ngamatea.

During the 1981–2 drought in Hawke's Bay Ngamatea took 1,000 cows from Matapiro station on agistment – some 22 truck-and-trailer loads of them. They were in a couple of paddocks up near the top, not far in from the main road, and the grass was so good it still was not chewed out after three months. The money from that grazing went to buy the rotatiller and a three-axle truck-trailer Richard also got them to buy. 'I towed it behind the big tractor in the winter, and I'd cart metal – they had metal quarries – so I'd pour it on the tracks, or in the gateways. I used to cart the bulldozer on the big trailer too, save walking it all round the station.' They also piled fencing gear onto it with the front-end loader and used it to lay fence lines, one man on the back throwing the posts and wire off, the other driving. Eric Brooking would come along and do all the lines he could manage, but there were so many miles of fencing he could not do it all and they sometimes had to use contractors as well. Twelve miles of new fence were put in in 1982 alone.

> We kept busy all the time – agricultural work, fence lines, metalling. Nobody had ever been over the country I was developing – just horses, horse tracks going out to Golden Hills. I've never had job satisfaction like I had developing Ngamatea and never will again; no one can ever take it away from me. All the area I developed, *I* developed. No one else did.
>
> Ngamatea used to be the Air Force's playground and they'd see you working away there from miles away and swoop down and 'bomb' you up, and you'd get a helluva fright. You wouldn't hear them – you'd just see the shadow go over the ground and think, Gee, that's a big seagull! I remember, it must have been the entire fleet, went over. They were flying below you, in the gullies.[30]

The station had CB radios by the time Richard started and he had been a CB buff down in Hawke's Bay. 'It didn't take me long to get the system upgraded quite a bit because it wasn't going that well. Then I sold my base set and aerial to the station, and friends came up and gave it a real tune-up. Today they've got radio telephones, which are so much better again.'

At the end of October 1981 Ngamatea was looking for a cook and the Mahoneys contacted Sheila Duley, whom they had known at Mangaohane. Sheila, who was down in Masterton, was very happy to come back to the high country – and to her friend Gordon Maxwell, still at Mangaohane.

Development in full swing.

He took over the mechanic's job at Ngamatea in the New Year, and they were married on the station in March 1982. It was a splendid occasion. 'Margaret said it seemed strange coming to the station all dressed up; this was the first wedding they'd ever had here.' The bridal cars were two big John Deere tractors, all decked out with white ribbons.

> We had a marquee – the boys helped put it up. I think there were 60 guests, and most of them stayed in the shearers' quarters for the night. People from town did the catering and waiting – in the shearers' dining room. I cooked some of the meats and things. Gordon's family came from Wairarapa and mine from Gisborne, Margaret and Terry came up, Elaine and Phil Mahoney were there and Eric and Hazel Brooking, Richard Whittington, old Pete Sisam, and the four shepherds. They asked what they should wear and I said, 'What do you normally wear to a wedding? Suits, or good sports clothes, but no jeans.' And not one boy wore jeans. They all either hired suits or wore good sports clothes – they were marvellous.[31]

Both Sheila and Gordon are deeply attached to Ngamatea. 'It's been home to us. We both come from rural backgrounds, we take this place in our stride. We've grown with it.' Gordon actually grew up with it too, in stories: his father knew George Everett at Te Awaiti and Ike Robin and his gang shore there 'and used to talk to Dad about coming up here to Ngamatea'. Gordon did his apprenticeship as a petrol mechanic, then moved increasingly into diesel. Like D'Arcy Fernandez and Alan Kennett, he has Chathams connections: he spent three and a half years there while

Spreading lime.

they were doing up the meat works, installing the generator plants and doing other associated work.

> When I was first here my main job was basically keeping the machinery going. There was no contracting anything out then – apart from now and again the local bulldozing contractor would come in and rip some metal from our only quarry at that stage, down below the woolshed in a paddock called Stone Quarry. Rotten rock – brilliant for farm use, it just breaks down well. Thousands and thousands of metres have come out of that, even since I've been here. It was a big sloping ridge when I came, and now it's a big dished out hole – and that's all spread around the farm, in gateways. I'd work with Richard and we'd do the metalling in the off-season when the Deere wasn't doing development. The station roads were done initially in that way – there were no contractors at all came on the place. I bulldozed some of the side roads further out – the bottom super bin, one going out to Trig bridge, and metalled most of it. Some of the bricks from the bread oven that was at the old cookhouse were used as fill in the swampy part that goes across the track to the Trig bridge.

> Gordon took over some of the bulldozing from Richard and put in dams and crossings. Initially they used concrete pipes, then asbestos pipes which were easier to handle but later recognised as dangerous, and finally safe, light-weight plastic pipes. With the development work, laying fence lines and vehicle maintenance, he was kept busy. Then there were truckloads of

material arriving at the station during the height of development, a lot of it building material from Rotorua carted by a firm from Rerewhakaitu.

> They'd leave at some ungodly hour of the morning to dodge the cops – overweight loads and whatever. And at that stage of it the hill either side of the Rangitikei bridge wasn't sealed as it is now. Quite often one of us would have to go down with the pivot-steer tractor and tow them or push them up the hill – they couldn't get up with the loads they had! Then they'd blow drive shafts because of the tremendous force coming up that S-bend on the station road. I think that happened a couple of times – they just completely screwed the drive shafts on the trucks – looked like a piece of twisted liquorice. That was a cost to them, not us! We'd always get them this far – we would go out and pull them up the S-bend and spare their drive shaft.[32]

When Gordon arrived the old cookhouse was in a heap ready to be burned. His first job, in January 1982, was to bulldoze the site for the new quarters. In 1983 a start was made on replacing the old ten-stand woolshed with a new eighteen-stand shed. It was done in two phases. There had never been covered yards so initially they just took down the old yards and built covered yards. Early in 1984, after the main shear, they started to demolish the old shed. The Woolaway contractors, who had done the covered yards, took the roofing iron off and the station staff collapsed the framework with the bulldozer. It took a power of pushing down. The woolroom, with Tom Whimp's extension, was all tied together with great lengths of No. 8 wire twisted into cables. They thought the whole thing would collapse when they got the wire loose, but were amazed at how strong it was. Eventually the old Studholme woolshed beams were exposed, with T. Waho, Hekenui Riha, Te Hata and other barely visible names carved into them.[33] The lot was unceremoniously bulldozed into a heap and burnt and the new shed rose from the ashes.

The station homestead was as busy as the farm. Elaine had four children of her own learning by correspondence and added Edwin Duley, Sheila Maxwell's son, to the class. She would be up at 4 a.m. preparing food, then wake the children at 6.30 and feed them and make lunches and away they would all go 'to give a hand' outside. For the school records this was called biology. They would be back early afternoon to do formal schoolwork while Elaine prepared the dinner. They turned the gardener's cottage into a schoolroom and the children tried to be seated and looking studious when the visiting school teacher arrived for her periodic visits. Elaine's teaching must have been good – all the children did well, including Edwin, who had had several changes of school. In May 1982 they had an unexpected addition to the curriculum: film-making.

The village of 'Te Puna' built at the southern end of the Swamp for the filming of Utu.

The little village of 'Te Puna' was built at the southern end of the Swamp for the filming of *Utu*. The film crew had found just the site they needed – a vast expanse of tussock and no sign of civilisation; the development of that area had to go on hold for a while. There were caravans and film people everywhere – director Geoff Murphy, actors Bruno Lawrence and Zac Wallace and swarms of support staff. They built sets and filmed out in the tussock, in the woolshed, in the shearers' quarters. Eric Brooking's new house had the best TV reception on the station so they went up there to play back their videos. The children kept popping into the make-up tent to see the magic: instant moko. 'One of the actors sat down with the kids and gave them an interview as part of their correspondence work. It was a great education for them. These were things which you never got to see at school.' The filming went on for several weeks and the weather was getting colder. On 14 May six or seven busloads of schoolchildren from Hawke's Bay arrived to be part of the march of the 'British troops'. 'We'd had a terrific row of frosts and fortunately it was perfectly clear all day. By the time they got them all back on the buses to go at 5 p.m. it was something like -1°. When you looked across at Peter's Range and the Kaimanawas you could actually see the frost glittering. It was extraordinary. It was just as well there wasn't a breath of wind that day otherwise they would have frozen to death.'[34]

Elaine alerted Nat and D'Arcy Fernandez to the excitement so they 'tootled up' and stayed a couple of days. 'They built all sorts of things

Breaking in the Swamp block, 1982; the Tit (far left) and Pinnacles (centre right) on the skyline.

out in the tussock – supposed to be houses, and they were just a front. There was a street and they'd tie the horses up. It was all make-believe stuff. And there's Bruno Lawrence in his four-wheel drive vehicle, roaming the tussock. They'd hired a lot of old muskets from Napier Museum or somewhere and they'd dropped them in the tussock and of course old Bruno had to drive over them.'[35] As soon as the filming was over the set was demolished and the station could get on with developing that block. On 30 January 1983 the world premiere of *Utu* took place in Hastings and all the station folk were given free tickets.

There was greater excitement, and consternation, to come. Just days later, on 4 February 1983, Ohinewairua station decided to have a burn-off in tinder-dry conditions on a very windy day. They had permission from the Council and a firebreak in place but the fire jumped that and spread down to the Rangitikei which proved no obstacle at all. It crossed the river just north of the Tikitiki bush and the next morning was spreading across Ngamatea towards the Taruarau and the Kaweka Forest. The Forest Service called in the Army and requisitioned bulldozers and drivers from every contractor in the vicinity; soon there were a dozen of them pouring in to the station from every direction. The Army arrived with their personnel carriers and field kitchens and set up camp in the horse paddock. A couple of hundred Forestry workers had established themselves in the big yard in front of the

Choppers called in to fight the fire, and the Army setting up camp.

workshop; the local county had a team on the job; farmers arrived from as far away as Crownthorpe – they could see the fire and didn't want it down their way; fifteen or seventeen helicopters with their monsoon buckets were parked in the paddock out by the old vegetable garden; the manager's house became battle headquarters, and all the maps were spread out on the kitchen table.[36]

The Army did the catering for everyone; it was said this was the best thing they did, although as soon as they arrived 'two nice gentleman appeared at the homestead to see if we could make some sandwiches for the troops while they set up their kitchen! And they didn't even bring us a bit of their tons of supplies to do it. But we made them – and they gave us a hand.' Communications were a major problem. The Forest Service was supposed to be in charge of the operation and they had their wireless in Elaine's kitchen; but the Army had their wireless down at their camp and the county people had their own communications. 'They were all communicating in and out, but not with each other. Trying to get the Forestry guys and the Army working together was nigh on impossible,' according to Elaine.

The helicopters were working from the station, coming in after dark all lit up – sometimes after looking for Army personnel kitted out in their camouflage and indistinguishable from their surroundings. The Army had big Iroquois helicopters and strict orders not to overfly 'sensitive'

M113 APCs assembled outside the workshop.

areas in case anything went wrong, but the chopper pilots from Taupo flew anywhere, including over the Army, with drums of fuel and anything else dangling from strops. The monsoon buckets were filled from the Ngaruroro and Rangitikei and from Lake Horotea in the Lake block. The bulldozers were cutting tracks and firebreaks through tussock and scrub up onto the Hogget, out to Golden Hills. The Woolwash bridge was still in place back then but half rotten and unusable. It was on the Army's map, though, so they went out that way and miraculously did not end up in the station's water supply. Kerry Henaghan, the head shepherd, and Richard Whittington were out there somewhere in the smoke with vehicle and bulldozer and CB radios, in touch with each other and the station, reporting outbreaks where personnel were needed.

The fire burned a strip right across the station – up from the Rangitikei, between the Tikitiki bush and Peter's bush but without penetrating either, and across the Swamp block and the Lake block. 'It jumped the Taruarau, crossed the Hogget and was going down the other side to the Ngaruroro when the wind changed and it sort of back-burned. If it hadn't been for that it would have been in Napier in a matter of hours because there was a very strong wind.' It burned parts of a new fence round a big block that had just been developed out towards Peter's from the Swamp. D'Arcy often had a working holiday on Ngamatea in the summer and the day before the fire he and Phil Mahoney had measured all 4½ miles of the new fence which in places crossed the scrubby little guts and gullies running out like fingers from the developed land. A few days later he and Pete Sisam went out to survey the damage.

> The fire had come down these guts, burned a lot of the posts, cooked all the wire – it was useless. It was only in patches on the top side, on the

The height of the blaze.

The fire advancing across Ngamatea.

Haymaking, 1950s.

undeveloped fingers – the new grass was okay. So we spent a couple of weeks repairing that fence. I think Phil sent in a claim to Ohinewairua, where the fire started, and they got some compensation for our time and the materials. I went down to the Rangitikei about a week later. There was soil, over the top of my boots, all cooked the heat was so intense. It had wiped everything out – there was just fluffy pumice. After the first rain they said the Rangitikei ran black. Then the Catchment Board oversowed with clover and grasses, all that face going down to the river, and of course the rabbits got in first. The fire only burned there for two or three days but there were hot spots left. I went up the Taruarau a week later, maybe more, and there was still smoke coming out of odd patches in the tussock country. Everything around it was burnt, so it couldn't go anywhere.

In the early days the musterers used to burn as they were coming in. Out at Golden Hills or the Boyd the tussock gets pretty high and they'd drop a match in it, but that was in the spring and it just burned the tops, it never burned the roots. But this fire burned everything. It was just like a desert out there. It's regenerated to some extent, but you can still see the effects of it 20 years later.[37]

The full fire-fighting team was on the station for about ten days, then a stand-by team of seven or eight was there for some time longer, with their

Haymaking, 1990s.

own field kitchen. The fire burned 14,380 hectares, 5,800 of which were on Ngamatea, and it was not declared out until 4 March. The station lost some fences and grazing, but no stock except feral deer.[38] Tree planting was subsidised after the fire and the station got a twelve-month extension to repay the existing development loans.[39]

Richard Whittington left the station later that year and had a few years driving his own truck and trailer, but with the end of the fertiliser subsidies in 1984 work had fallen away; and when his replacement at Ngamatea, Andy Lysaght, left, there was a job for him again. He had married in the meantime and in 1987 he and Karen and their baby son moved into the former single-men's quarters, now the casual quarters, and were there for the next two and a half or three years.

> That second time around I'd bought a round hay-bailer. The contractor used to come out from Taihape but he might get out there late April because he'd done all the Taihape farmers first. The station had a good hay-mower which they'd bought basically to help out the contractor mowing the hay, and for topping thistles and that. So if they bought a new hay-rake and I bought the bailer I could use the station tractor and we'd do it in February, in good time. We got good quality hay, made when it suited us – and another job, in my spare time, was stack all the hay away in the barns. We

Aerial view of the Ngamatea basin in 2000, looking towards the Otupae range; the station is in the middle distance.

used that big trailer. So you kept pretty busy with the development and the cropping and the hay.⁴⁰

Development of part of the old Swamp holding paddock was begun in 1978 before Ray Birdsall left and it was sown in March 1979; Richard Whittington worked up the last part of it to be developed after the filming of *Utu* and sowed it in grass. Development at Ngamatea continued until 1986 when Andy Lysaght put Sunny Face, the northern, lower end of the Road block into grass. The Ngamatea Swamp remains undeveloped and is a Recommended Area for Protection under the Protected Natural Areas Programme; as well as significant landforms it has many rare and endangered plants. The period of intensive development at Ngamatea is over. Of the station's 28,300 hectares, 8,200 are now in developed grass and another 10,700 are grazed oversown tussock grasslands which will not be developed through cultivation. There has been extensive consultation with soil scientists and botanists and care is taken to fence off sensitive country in riparian areas and incorporate them in a fertiliser management plan. Initiatives such as these help to protect the catchment by leaving wide buffer areas which drain into the major rivers.

Chapter Eight

Cooks, Cops and other Characters

We remember the good times and forget the bad.

Everyone who has ever eaten in station cookhouses is full of cook stories – good cooks, terrible cooks, those who stayed for years and those who stayed one day. 'Cooks!' said Bruce Atchison, 'if you swotted them all up and studied them you could just about write a book on the cooks alone.' Few of them were there for the right reason, some made home brew from the yeast they were supposed to use for bread, some drank the meths they were supposed to fill the lamp with, and a day or two off the station was the downfall of many. One must have arrived with no money: his fare from Taihape was paid by the station and taken out of his wages. When he disappeared two and a half months later a note in the wages book explained it all: 'Did a flit. Police inquiring.'

Every station needs a handy fellow who can cook at a moment's notice, and Colin Kirkpatrick, the great gardener, was such a one at Ngamatea. He was there for years and got plenty of practice. One cook had come to the station with his mate, a general hand known as Jack the Ripper, and after a few weeks of great cooking he up and left and Colin took over. It was the off-season, so he had only a couple of shepherds to cook for. A week or two later the old cook reappeared in a taxi when Lawrence, Peter Cameron and Ivan Chapman were drafting cattle in the yards. 'We thought, Ah, here's our good cook back, but he obviously wanted to talk to the boss and he kept the taxi hanging around until we'd finished drafting.' When he asked if he could have his job back, Lawrence said, 'What job? I don't want you back', and turned and rode off. 'Well, he just got back in his taxi and drove off – all the way back to Taihape!'[1]

One new cook, a big wild-looking fellow, put the porridge out the first morning and while everyone was busy eating they suddenly heard brrrrr, brrrr! He had taken a great butcher's knife and driven it down between two boards on the slide, then pulled it back and let it go to make a noise that reverberated through the dining room. When everyone looked up startled, he announced, 'I'm the new cook around here, and what I say goes.' His reign was a short one. Even shorter was that of a cook who arrived on the

Friday on the mail truck, with his bicycle. Colin had been doing the cooking again and he had the evening meal ready. Next day the new cook made breakfast and lunch, and in the afternoon when the boys were working in the yards they saw him pedalling along the track on his bike. 'Oh, he must be going to have a look around, we thought. Never to be seen again. Gone. That was the shortest cook we ever had.'[2] But then there was Jack Dumbelton, who must have been a station cook for the right reasons: he stayed from 1936 to 1945 and broke the record for longevity on the job.

There was a cook who fancied himself as a boxer. He would put a few mattresses down in the dining room, don his gloves and give the boys a bit of instruction in the art of knocking people silly. He got at George Everett one evening: 'Come on, George, put the gloves on and have a go.' George declined politely: 'I don't want to hurt anybody'. 'Oh, come on, George, you won't hurt anybody – put 'em on.' Night after night the cook kept teasing until George did put the gloves on. 'And of course this cook comes out like Jack Johnston, dancing all round George and lets rip and hit George a couple of times and then George just went Boom. Bang, down went the cook. One bang and he was gone.'[3]

> One of Lawrence's sayings was 'The Good Lord sends the tucker, and the Devil sends the cooks'. We had some brilliant cooks, and others – you wouldn't want to eat there. There was one Jimmy Hughes – he was on the transcontinental railways, across the States or Canada. He was a top-class cook – could have got a job in any classy hotel anywhere. Don't know if he was English or what he was, but he was a booze artist. That was his downfall. He used to bake all our bread in the big baker's oven down at the cookhouse – made the bread for the whole station. He'd cook anything – pastries, top-class meals, but he'd come out and he'd have his carving knife in his hand and he'd stand at the head of the long table they had in the cookhouse and he'd say, 'Any complaints?' – and no one was game to say a word. He was there for a fair while but he'd go to Taihape every now and then and I'd have to go in and bring him home – go round all the pubs to find him. That was the downfall of a lot of the people out there.[4]

One cook used to come back by taxi from his binges in Taihape. One day, feeling mellow and thinking what a good fellow the taxi driver was, he decided to reward him with a mutton from the big walk-in chiller – there would often be three or four hanging in there. But Lawrence was awake to the tricks and next time the cook drew his pay he found the price of one mutton deducted from it.

> We had about five cooks that first year at the station. One of them was a woman – and she was a big lady this one – and there was only the one

bathroom. She's lying in the bath there and one of the boys is just coming in for a wash and hello, the door's locked, so he knocked the door down – and there she is. Well, she out of there and she up to the boss. We could hear her going across the paddock, bellowing and roaring, and the boss says, 'I'm going to town on Friday, I want you off the place by then.'[5]

When Ray Birdsall was head musterer cooks were the bane of his life:

You see you took 30 loaves of bread and all your stores out to Golden Hills and after a fortnight your horses were worn out, so they had to come back in. The packman would bring them and usually one man with him, and you got another 30 leaves of bread and more stores to head for the Kaimanawa side of the station. It would be arranged that on such and such a day I wanted 30 loaves of bread ready to go, and you'd come in with so many packhorses and all the hacks and that day all the rest of the boys would be mustering on foot. Well, you'd get in there and the cook would be drunk as a lord, flaked out, no bread. And I'd have to threaten that I was going to beat him up if he didn't get out of bed, and I'd make him work all night because I wanted to get away next morning with 30 loaves of bread. I always managed to do it, but it was hard work – my godfather it was hard work.[6]

Pohokura got more than its share of odd cooks – or maybe the remoteness just brought out the eccentricity in them. There were often only two people down there, the cook and one shepherd, but during the muster it would be a bit crowded with three or four musterers bunked down in the dining room of what was really a two-roomed whare with a lean-to kitchen at the back. That cookhouse had been there since the days when Pohokura was an outstation of Mangaohane. Lawrence extended the accommodation with a couple of prefabs out in front of the cookhouse and when Jack started back there he slept in the end room of one of the huts. They would be in their bunks early because of the 3 or 4 a.m. start.

One night I was lying in my bunk and I could hear this slide, click, snick, slide, click, snick, and I couldn't figure out what was going on. From my room I could look into the cook's room over in the cottage and there he was, sitting on his bed playing with an old .303 rifle we kept down there in case you had to put a horse or cattle beast out of its misery. I sneaked out of my room and woke the other boys and we were all hiding in this one room. Everybody had a bit of a rouse around to make sure there was no ammunition lying about he could get his hands on. The first opportunity we got that rifle. You wouldn't know what the hell he was going to do with it.

Then there was a cook who went 'moon mad'.

> It was well into autumn and it was getting near to the full moon but the nights were pretty cloudy, and about five one night Dave Bower rang me. He was a great friend of my father's and he used to come back from time to time when we were short-handed or there was no one at Pohokura. I knew from the way he was speaking that the cook was actually listening and all he said was, 'I'd better bring this cook up.' I said, 'Oh, leave it Dave and come up in the morning', and he said, 'Nah, I'd better bring him up tonight.' I wasn't there to meet him – they had to ride up from Pohokura and when they got down into the yards at Timahanga here, the boys saw Dave had a rifle across his saddle. Dave hadn't carried a rifle since Gallipoli days. The cook was in front on the quietest old hack from down there, and Dave on his horse with this rifle, and the boys said, 'What's the story with the rifle, Dave?' And Dave said, 'Oh, I just thought I might have seen a deer.' Well, you couldn't see your hand in front of you. The cook parked up at the cookhouse down here, and as the night went on he started to get more and more agitated. He hadn't had any booze, and it finished up he was chucking furniture around in the cookhouse and beating his head up against the wall and sitting outside howling like a dog, and the boys didn't go to bed. I didn't know anything about this till I went down in the morning and everybody's red-eyed, and I said, 'What the hell's going on?' And they said, 'It's the cook – he's moon mad.'[7]

The greatest character of a cook was Leo McSweeney who became a permanent fixture at Pohokura after doing part of the 1955–6 season as packman-cook on the station. He came from a well-known and highly respected Taihape family but he was the black sheep of the clan and he kept his distance from them. The boys reckoned he was the best station cook they had come across, and what's more he was good-tempered and had a great sense of humour. One of his little habits irked them, though: he made their toast on the hotplate of the little Orion wood stove, and to test whether it was hot enough he would spit on it. 'If the amount of sizzle was right the bread would go straight on top. No one dared not eat it! But he made lovely roast meals – and breakfast at docking time was mountain oysters, always a real treat. He had an excellent recipe – a brown sauce and onions and he'd just braise them to perfection, and we had them on our toast.'[8]

For the most part Leo's only company at Pohokura was the single shepherd who spent the winter down there; Paul Thomas had six months with him in 1959.

> When I arrived down there on the first night I think I had one of the nicest meals I've ever had. It was roast duck and gravy with all the trimmings,

vegetables all grown on the property. Leo kept his own garden, and he used to catch the ducks with a bit of bread on a fish-hook – no problem to him. And his pudding – if you didn't bite hard on a bit of baking soda he hadn't ground up properly, it was a pretty good pudding – big steamed pudding. He was an excellent cook, and he looked after himself very well, and by doing that he looked after me too.

He was an interesting sort of fellow – very different. For goodness sake, you put a 65-year-old and a 21-year-old together, and one's as Irish as Paddy's pig, and a Catholic at that of course – he was at the exact opposite end of the scale from me. I've probably had a slightly privileged upbringing, and Protestant. We couldn't have been further apart, but we did get on reasonably well together. At times I'd enjoy his company, at times I'd get fed up and just leave him to it. But I don't think we ever had an argument or got totally upset with each other. He never worried me and I hope I never worried him, but he did have a problem, without a doubt – like a lot of the cooks. I guess that's why he was back there.

It wasn't easy for him either. He kept a very clean cookhouse really – and he had to winch the water up from the creek. There was no running water there. There was a windlass from just outside the cookhouse down to the creek and you let the bucket go and it flew down the wire and filled itself at the bottom and then you wound it up to the top. *Beautiful* water. Best water in New Zealand. It came out of the hill, not very far from where the windlass went in, just bubbled out. Probably why the hut was put there. Then of course, with no electricity the water was heated by the wood range – no cylinder, no wet back – just a part of the range, a tap on the side. Oven on one side of the fire-box and water tank on the other. That was the only hot water.

He didn't do any stock work but he'd get on a horse occasionally. Well, he used to have to ride down to get his ducks from the dam in the bull paddock! I don't think he'd ride to Timahanga though. If the Land Rover didn't go up I don't think he'd go up. He cooked and gardened, and he reared pheasants and had his chooks and turkeys. There was no cow to milk – we had milk powder. And lighting of course was tilly lamps, which forever seemed to need new generators. They gave no end of trouble.[9]

Jack knew Leo well, and he continued the story:

Old Leo died down there round Christmas time in 1964. Too many people had given him grog for Christmas and he'd cleaned the lot up. Peter Hill who was general hand on the station was from England and had nowhere to go for Christmas so he was staying down there with Leo. He'd gone fishing and when he got back he found Leo was dead. He managed to get through on the

Dave Bower and Leo McSweeney at Pohokura cookhouse.

old phone to us at Timahanga. Mum and Dad and Missy were at Ngamatea, and they got through to the police. Missy and Dad came down in the old short-wheel-based Land Rover, and the police arrived, and Dad and Missy stayed with Jenny and I took the two cops down to Pohokura. It would be about nine o'clock at night by the time we'd forded the rivers – luckily the streams and rivers were low at that time of year. We had to get old Leo out of his room and they wrapped him in blankets and a tarpaulin.

We started off from Pohokura with Leo sort of diagonally across the little well-shaped deck of the Land Rover and the three of us managed to get in the front. I was driving and the two cops beside me. Nobody sat on the back with Leo and it was pretty rough coming up the old Spiral; he was getting shaken around quite a bit. We got to Timahanga and the police just headed off in their vehicle, and Dad and Missy got into the Land Rover and headed up towards Ngamatea with Leo on the back. When they got to the end of the Strip the hearse was pulled up there on the side of the road to pick up the body. Missy said it was eerie, like something out of a ghost story. The mist was swirling around in the part moonlight and the undertaker was in his pin-stripped trousers and tail coat standing rigidly to attention beside the hearse as they drew up. It was like a Hitchcock movie. They helped load Leo into the coffin and into the hearse – no cops there to give a hand – they'd gone back to town to resume duty, and the hearse went off to Taihape and Dad and Missy drove back into Ngamatea.

There was a post mortem and we went in to the funeral two or three days later; he was buried in the Taihape cemetery. Some of Leo's family were there, Mum and Dad were there and Jenny and I, but most of the station people were still away. I remember two distinct sounds – one when the coffin hit the bottom of the grave, and the other as the reverend gentleman who took the burial service shut up the Bible and all his good books. Then he walked straight round the grave to me and said 'What's the fishing like in the Rangitikei these days?' I'll never forget it. Round the open grave everyone had a discussion about what the fishing was like, and politics and the state of the country.[10]

The other great characters who were at Pohokura for years were the Harkers, Old Tom and Young Tom. Young Tom started in August 1948 and his father joined him a year later. Officially they were shepherds, but Young Tom did a couple of seasons mustering as well, and Old Tom could turn his hand to almost anything. There was a cook at Timahanga, but the Harkers looked after themselves down at Pohokura. Old Tom had only one good arm. He had had an accident in the bush when a wire cable snapped and ripped through the back of his left shoulder and he was left with no movement in that arm, but it did not stop him riding, crutching, driving, breaking horses. Lawrence had bought an old four by four Chev from the Army, mainly because it was so high it could be driven up and down Jimmy's Creek – part of the old road to Pohokura. Jack remembered him driving it. 'He'd jam the steering wheel with his knees and reach right over and change gears with his right hand. He had his own vehicle too – I don't know if he ever had a licence!'

When Don Hammond arrived to start work at Ngamatea he was surprised to be dropped off at Timahanga. After unpacking his gear, he was allotted a horse and told that he and Tom Harker would be taking a mob of cattle up to Ngamatea next morning.

> I saw this old guy there trying to saddle a horse. I didn't know who he was, and I offered to help him. Gawd, that was the worse thing I ever did. He was really up in arms about anyone trying to help him. I never ever offered again. He managed pretty well, used to go into Taihape and drove wherever he wanted.
>
> Anyway next morning we had to go down to Pohokura and pick up this mob of cattle. We got past Jimmy's, riding along through the scrub and all of a sudden he lets out a scream and a yell and kicks his horse and his dogs take off

and they pull down a beast. Of course my dogs all took off too, and I thought, What the hell's going on here. He jumped off his horse and yelled at me to get a hold of it – it was a cleanskin – a station one that had never been earmarked. He pulled some thongs out of his pocket and he tied its legs together and left it there. It happened three times on the way down – we had three wild cattle tied up on this track, down this gully. We went on and picked up the mob at Pohokura and drove them back and as we came to these ones we'd tied up, we mobbed the cattle round them and let them go, didn't mark them or anything. They were station cattle, scrub cattle that had missed being marked, and once they'd been tied up and dogged they mixed in with the mob and walked out with them.

So that was my introduction to Tom Harker. I wondered what the hell I'd struck. But he was a great yarn teller, and he was always talking about the pigs he'd hunted and how long their snouts were. He was a real character, but he used to tell some terrible tall stories – you had to be dreaming to swallow them. Then his son, Young Tom, he was a big gangly guy, only about nineteen I suppose. He was full of all the bull that his father was, but he finished up in our gang as a musterer, the year I was there. He had a terrible temper. He was a shocker – pretty hard to live with, especially in a small hut with six other guys.

When Bill Cummings was head musterer and Ash Watt second musterer there was a bit of a drama one morning in one of the huts.

I was outside when it happened. I don't know what Ash said, but Ash wouldn't say anything to anyone that would offend them. But this young Tom Harker reckoned Ash said something and he threw his meal in Ash's face. I just walked into the hut as Ash said, 'Come out here, Tom, I want to talk to you.' I didn't know what had happened and when Ash took him out I followed them. Ash said, 'Tom, I don't advise you to ever do that again. You'll get yourself into big trouble.' And that was Ash. He could have eaten this fellow – but that's the way he handled him. And he respected Ash after that. Nothing more was said, ever. That's the way he treated men, Ash. Great leader.[11]

According to the old roadman, Bill Wells, Young Tom was a great mimic. 'He would ring up the head shepherd on the station phone, pretending he was Lawrence and he'd say, "Ccom muup and mususter the Guully block." Old Tom was quite a respectful chap actually. He used the rifle with one arm, he could shoot pigs – a wonderful shot.' He used to do the packing too, between Timahanga and Pohokura, and was 'an amazing horseman'. But they were a scruffy pair, real backwoodsmen. As Jack remembered, neither of them ever wore socks in their boots. 'We used to break a pair of boots in by filling the boot up with neatsfoot oil, and they

Old Tom Harker.

used to fill their boots up and jump into them with no socks and break their boots in. But the problem was after they'd done that, they never used to wash their feet.'

Gordon and Beryl Grant were neighbours of the Harkers for a couple of years when they first went to Pohokura. Once they were on the way home after a particularly slow trip from Hawke's Bay and when they got to the Harkers' Beryl was starving. She was a great housekeeper herself but she knew how dirty the Harkers were.

Old Tom says, 'Would you like something to eat, Beryl?' And Young Tom's sitting there, rough as guts, his hat on, and Beryl says, 'No thanks.' 'Do have something – you must be hungry.' And Beryl spots these eggs sitting on the table, and the eggs have got chook poop all over them and feathers stuck to them, and she thinks, 'I can't come to much harm with them, I'll ask him if I can have a couple of boiled eggs – they'll get clean as they're boiling.' So she has these boiled eggs, but she must have lost her sense or something, because he says, 'Would you like a cup of tea now?' and she says, 'I'd love one.' And he made the cup of tea with the water he'd boiled the eggs in! And she had to drink it. She laughed about it for years afterwards.

Old Tom would bring the mail in to us, once a week, and bring a mutton and the groceries or whatever you wanted from Timahanga. He'd ride up and take a packhorse. Good people – they suited that place in those days – the type of people you wanted.[12]

There were horses running almost wild on several blocks at Ngamatea and Timahanga and the Harkers knew how to handle half-wild horses, and what was needed for working in the scrub down their way. Each year a couple of musterers would be sent down to do the shearing muster. Ray Birdsall remembered Lawrence sent Bruce Atchison and him the first year Ray was there. They rode the 20 or so miles down to Pohokura on the hacks they'd been using all season, with their packs on the back of their saddles, and Old Tom shook his head. 'Those horses won't be any good down here – they won't go into the scrub good enough', and away he went and rounded up a mob of horses and chose a couple that suited his idea of what was needed to do the job.

The damn things had never been broken in properly – they'd just charge off into the scrub, they'd buck all the time, and talk about a picnic. It was terrible. Bruce and I had to drive mobs from Pohokura up to Timahanga every day and bring a shorn mob back. We were starting at half past two, three o'clock every morning and we'd muster them down to that little swing bridge over the Taruarau River. It was only one sheep wide and you might have 1,000 ewes and 1,000 lambs and you'd hold the mob there and get them started across and suddenly one would decide to jump over and turn the wrong way, and you had to climb out right over the bridge and turn it round. We got about half the mob across this day and suddenly they weren't coming off the other side, so we had a look and the boards had dropped off the bottom of the swing bridge and they were all going down like a dip into the river. Once they started that we had to put the rest in the river and swim them across.

This is in January – hot as anything – and the sheep just wouldn't move. We were punching them as hard as we could go – and we had plenty of good dogs, but we didn't arrive till after five. We had to turn around and take a mob back to Pohokura. Mr Roberts said, 'If it gets too dark you'd better leave them at Jimmy's for the night.' We were going over the top of the Spiral and suddenly we heard dogs barking in front of us. We thought, That's funny – what's going on. And it was Old Tom Harker. He'd decided on his own bat to bring another mob up and we hit bang together on top of the hill. We had about 1,000 ewes and their lambs and he was bringing a small mob up. It was pitch black – couldn't see a thing. There was a real row over that lot.[13]

When they wanted to get more horses in they would chase them down from the Otupae Range and the Harkers would catch them with their ingenious snare – big bundles of scrub about 5 feet high with a heavy rope

Repairing the HD9 in the scrub; George Everett and Gordon Grant listening to the fourth test against the British Isles in September 1959, with the aid of the battery from the tractor and a flagon or two.

tied through them. They would put these bundles on the tracks where the horses raced down through the scrub and the rope looped across the track would catch the first one that came down and he would have to tow the bundle until it pulled him up.

> They had a horse called Mudguts with a big Mexican saddle on it, a fantastic roping horse. You'd lasso a bull or an unbroken horse and as the pressure came on the front of the saddle Mudguts would turn and brace himself and hold them.
>
> One time we had this mob of horses, half a dozen or so of them, in the Gully block and Road block – all the fences were down and they were all in together, and when Jack Cornwall became head musterer he decided he'd round them up and break them in. They were about seven-year-olds, unhandled, and it took us several days to run them down but eventually we got them into the yards and it was the Harkers' job to break them in. They used to spend one day on each horse and then they'd give them to you – and buck – you've never seen anything like it.[14]

When Lindsay McRae came back to break in horses after he had done a season's mustering in 1954–5 the numbers had built up. In his first year there were about 30 horses averaging nine years old. Most had never been touched, but some had been partly broken, then turned out again. He put in long days each year for about five years, handling three horses at a time, for about a week.

> Three would fill in a day nicely – you can only do so much on a horse at a time, or it goes sour. Then as the next three got done, the first three would come back in for a day and a half, two days, and be gone right over again, and ridden. They were going to be culled and some of them sold. I finished that first year and eight or nine weeks later old Lawrence rang me and instead of saying could I come up for a day and go through them and take out the good ones, he said, 'Could you go down to Stortford? We're sending the sale mob down.' Unfortunate, because two-thirds of the those that came down were the good horses. However, that was his way of doing things. I think sixteen came down and were sold at auction. If I'd been up there I'd have altered which way they went. I went down and spent a day with them before the sale, and rode them at the sale.
>
> From there the mobs were down to about eight or ten a year so I'd only be there for about three weeks, in the latter half of January, February. The sale wasn't until early winter, and they weren't touched again up there. Even when they were drafted, they were just chased around the yard and chased onto the trucks and never had a hand on them till I took them off the trucks. One good one sold well, but the rest didn't really. About £5 would stop them and even in those days they should have gone for £15.[15]

Reg Gillan, the station hand known as Cactus because he liked to gallop around on the children's pony making believe he was a Texan cowboy, was only a little fellow but he looked the part in his big hat with rabbit traps dangling either side of his saddle. He had no sense of direction, and as he rode out from the station he would tie knots in odd clumps of tussock so he could find his way home again. Of course the boys would go out and untie the knots and tie others in quite the wrong direction so poor old Cactus would get lost and arrive home late for tea. At mealtimes he would have the boys boggle-eyed with stories of his time as doorkeeper of Auckland's biggest brothel before he forsook the bright lights for the glow of the tussock. 'These stories he told,' said Winnie, 'well, I suppose they were cookhouse stories. Sometimes he would ask if he could come in and play the piano – and he was a really good pianist. He liked to play his signature tune which he'd composed himself. It was called the "Brothel Jangle"!'[16]

Wonderfully eccentric old Bert Jeffery was one of Ngamatea's great characters. He admired and respected Winnie Roberts and always had a poem to recite to her. He thought everything in rhyme and she gathered quite a collection of his poems recording all sorts of happenings on the station.

Jeff the poet.

Jeff came in one day as drunk as a lord. He knew just how much we loved our land and everything about it and he said he'd written a poem. He could hardly stand up. He held onto the table for support and I said to him, 'Well, tell it to me'. He never faltered – he could just repeat it. And after he'd finished I said 'Sometime, Jeff, write that out. I would like that.' He used to bring me all his little poems and things, so he wrote this one for me and I've kept it all these years, and I think a lot of us feel the same way about our land. It's called 'No Sale' – and I can see him now, wobbling and hanging onto the table to keep his feet while he remembered it.[17]

> *No Sale*
> So you want to see my farm he said
> To see my farm said he
> Well there it lies before you now
> With fallow rich from the turning plough
> And the living green of the almond bough
> And the homestead warm on the hill.
> It isn't much of a farm he said
> Not much of a farm said he
> But the living dream of forty years

And a plain old tale of hopes and fears
Of a man's hard sweat and a woman's tears
Are something to think of still.

It's just a bit of a farm he said
A bit of a farm said he
With the young wheat green in the morning glow
And the red cows there on the flat below
And the blue smoke hanging pale and low
And the swallows over the sward.
It isn't much of a farm he said,
Not much of a farm said he
But a man's own plan is a precious thing
Thrice rich to commoner as to King
And he is rich whom the seasons bring
The prize of his toil's reward.

So you would inspect my farm he said
You'd look at my farm said he
Well there are the fences straight and good
And the homestead built of the native wood
And the cleared land where the scrub once stood
And the rest is as you see.
It isn't much of a farm he said
Not much of a farm said he
But such as it is it's all my own
Built of my faith and blood and bone
From the first hard toil to the crop new sown
And the grandchild at my knee.

So you want to buy my farm he said
You'd buy my farm said he
Well how do you value the light and the shade?
What is your price for the dreams I've made?
And how would you buy on size or grade
The children whose shouts you hear?
You haven't the money to buy he said
This bit of a farm said he
You haven't the money to buy the worth
Of the joy and prayer, of the death and birth
The power that blessed this fruitful earth
And the love that made it dear.

                              A. J. AT NGAMATEA

On the lawn at Ngamatea. Back row: Dr Bathgate (right) and three sons; front row: Jack Roberts, Winnie Roberts, Lester Masters.

Dr Bathgate and Lester Masters were two early visitors at Ngamatea. Both were pioneering types, keen trampers and conservationists who happily made themselves at home in huts on any part of the station – especially if there was a resident rabbiter who was a dab hand at scones or camp-oven bread. Dr Bathgate recorded some of his adventures in a series of articles in the local press. His first visit was to Pohokura for the shooting and fishing and a lie in the hot pool. The old cook down there seemed happy to host a party of three for a few days and when they left, loaded up with venison and trout, he added a couple of wild lambs he had killed for them. All they needed was a packhorse to cart it all the ten or so miles up to Timahanga where they had left their car, a big open Buick. No problem – the cook could provide a packhorse too; the boss would be glad to have them take it back up to Timahanga for him. Susie turned out to be practically unbroken and she took off through the scrub heading for home, shedding bits of the load as she went. The shearers caught her at Timahanga and unloaded the pack-boxes which survived the journey, but all their personal gear, stuffed into sacks and heaved up on top of the load, had been scattered over the countryside. When they staggered in through the rain, loaded down with all the gear they had managed to glean on the way, the shearing gang gave them a rousing welcome and complimented them on their kindness to the packhorse by carrying half her load for her.

A few years later the intrepid Dr Bathgate went in to Pohokura in an ex-US Army jeep – the first vehicle ever to drive up to the door of the cookhouse. By this time there was a rough track bulldozed as far as Jimmy's, but after that it was the pack-track and a prayer. They were armed with shovels, spades, wire-strainer and axe and in one hairy spot on a sloping papa face with a drop of 30 or 40 feet to the creek below they had to dig a shallow trench for the top wheels, to stop the jeep sliding sideways down the slippery slope. They did a bit of navvying to get through a pumice and papa cutting into Jimmy's Creek, then bumped their way down the creekbed to the river and floundered through that up onto the grass and an easy ride to the cookhouse. The cook was so startled to see a vehicle at his door he rushed to the phone. 'Boss, boss, there's a party just arrived at my cookhouse in a Jap. Too right it's right, they're right here at my cookhouse door.'[18]

Pohokura and Ngamatea were Dr Bathgate's playground; he had become a friend of the family and 'sort of the family doctor', and was always welcome up there. Most years he and his son would be there for a week or two and if they got a few trout and a deer they were well satisfied. Ivan Chapman remembered Dr Bathgate used to borrow a packhorse:

> Old Royce was his name – a very quiet one – and they went out and stayed at the Boyd hut and Golden Hills and did a bit of fishing and a little bit of shooting. They came back in and Royce had rubbed a wee bit on the shoulder where the pack-strap came around the front and they put a great big bit of sticking plaster over it before they put him out in the paddock. We went to get the packhorses in a few days later and we're asking 'What's that on Royce?' and it's one of those sticking plasters with a pad on it, you know – a Bandaid.

Lester Masters knew the Ruahines well and had done a lot of pig hunting, and he had been a shearer in Australia. He was 'a tiny little joker about 5 foot nothing and he'd shear merino rams and they'd stand up taller than him'.[19] Bill Cummings and Ralph Atkins, coming in from the back, saw a little footprint on the pack-track. 'There's a woman out here,' said Bill. 'No,' said Ralph, 'that'll be Lester Masters – he's got little wee feet, and he walks out to Golden Hills.' Lester walked everywhere: he tramped through the ranges, stayed in all the huts and left visitors' books with one of his poems in each of them; he was a well-known bush poet.[20] 'He came into the stables one day, the old stables, one real little dinky-die guy, yakking away, and I bought one of the copies of *Tales of the Mails*. I've lost it,' lamented Dave Withers. 'I'd like another, if you ever find one. He talked to me about the stage coach, and getting the mail through.' He also contributed columns about the outback to the *Auckland Weekly News* and he would sit in the cookhouse at Ngamatea and get the musterers talking and record their idiom. He found Ivan Chapman sitting over a late meal one day: 'What's it like out the back? It's supposed to be pretty steep, isn't it?' and Ivan explained, 'Well, put it this way, what's not straight up and down is overhanging – and he went and wrote that in the *Auckland Weekly*!'

D'Arcy Fernandez knew Lester well. When he and Ash Watt were packing chaff out to the Boyd at the beginning of one mustering season, Lester wanted to go along. He had put a visitors' book in Golden Hills hut – 'in a tin folder so the rats wouldn't get at it' – and he wanted to put one in the Boyd hut.

> Ashley gave him the packman's hack which was used to following the packhorses, and it was a hard-mouthed animal – Popeye, I think it was. That was all right going out because the packhorses used to just steady plod all the way and we got out there and spent the night at Boyd's. Ashley and I had rifles just in case we fell over a deer, and coming back they put the rope round each packhorse's neck and turned them loose. They knew where the station was – and they took off. They didn't walk – they galloped home. And here's old Lester, on this horse, trying to hold it back, hold it back, hold it back, because we weren't in a hurry to go home, but the horses were. He was like a little jockey up there trying to rein his horse in, but in the end he gave up, it was taking too much out of him. We yelled out, 'Right-o, we'll catch up with you down the track' – and we never caught up to him till the yards at the Swamp fence. All the horses were there, waiting at the gate. Lester got there a long time before us. He was a character.[21]

Bill Sommerton mustered at Ngamatea for just on two years from the end of 1962. He was in his early thirties, older than most of the musterers, although Arthur McRae was there when Bill arrived, and he was just a year or two younger. According to Alan Bond,

> Bill was an unusual fellow – a helluva good fellow really. Bit of a mystery man – nobody knew where he came from. He arrived up there with half a dozen dogs it looked as if he'd picked up off the street, broken-down boots, a Donegal tweed sports coat, bush shirt and raincoat – that's about all he had. He was a dag. Dry – we got a lot of laughs out of him. But could he kill sheep. He was pretty handy with the knife so he was always the man for the dog tucker.

Bill turned up at the station when they were particularly short-handed; he had answered an ad in the paper on his way down from the north, picking up a team – well, a collection – of dogs, on the way. Dave Wedd and Arthur McRae were there and they had set out to muster the Hogget.

> We were coming in from the Rock to Log Cabin and this great big man walked in, big bush shirt on and trousers and these boots – the seams had gone and they were crusty and dry, and the young blokes just couldn't believe it. He introduced himself – here was the new shepherd just arrived. He'd left the bush away to hell up in North Auckland somewhere and he had about seven dogs, and they're all on chains, he's leading them tied to horses' tails, but the boots – you wouldn't believe. He got to the station somehow and the boss had sent him out to meet up with us. But he had a great way with him – he'd been around a bit. Damn good man.
>
> Anyway, next morning we were saying where to go – some would go back above Timahanga, well up, and you'd muster everything around towards the Taruarau. There were quite a few walking beats too, down some rockier country, and he said, 'Don't you give me a walking beat – no one could give me a walking beat with these boots.' So Bill and Arthur and I went one way and we got behind the stock. You get behind these big woollies – there'd be two or three, just in a little pocket, and you'd get them coming along, but one would go down and lie there, and the others would be drifting along and then you'd see them bleating and they'd come back – they'd know their mate wasn't there. So a lot of the time you'd cut their throat, and one big one just ran out onto these rocks, so I said to Bill, 'Oh, well, cut that ewe's throat'. I'm thinking he'll just knock it off and catch up with us, but Arthur and I are way, miles along, and no sign of Bill, so I went back, and here's Bill, sitting on a rock, the dogs are all eating – he'd chopped the sheep up – and he's munching away on a great big bit of fruit cake. Oh, he was a bottler, that bloke, he was a beauty.[22]

Jack Roberts remembered Bill had a 'little prick-eared bitch' who used to jump up to try to see over the tussock.

> Bill would call her, take his foot out of the stirrup and she'd jump onto his foot, race up his leg and sit on the front of the saddle, on his coat. When he wanted her to run he'd pick her up, point her in the direction of the sheep, and when he had her all lined up and she could see the sheep, her ears would prick up and he'd give her a bit of a nudge and away she'd go. When she'd finished her run he'd put his foot out, and up she'd come again.
>
> Old Bill, when he got a few beers in he'd want to dance – we called him the dancing bear. He used to have one of the biggest hacks to carry his weight – 16 or 17 stone. Down at Timahanga when Bill Jones' shearing gang was there and there was a bit of a party on, he got up and started to dance, then he grabbed one or two of the rouseabouts to dance with, but he had his hobnail boots on, size twelve or so, and next day about half Bill Jones' rousies couldn't work because he'd crippled them with his hobnail boots.

Packhorses crossing the Taruarau on the way to Golden Hills, 1953–54 season.

There were more horses than people on Ngamatea and some of them, especially packhorses, were characters in their own right. There were rough packhorses and amenable ones, some that took special loads and some that objected to taking any load: Kitty used to take the big pack-boxes with the bread and others would take only the boys' swags. Some always led the pack team while others would try to challenge and take over the lead. Some started out as hacks and ended as packhorses; some were loved by one musterer and hated by another. And, funnily enough, girls' names were thought to be highly suitable for packhorses, but boys' were not. So there were Bessie and Dolly and Winnie and Kate – but Cobber, Tarzan, Trooper and Gunner.

Ivan Chapman remembered that Winnie Roberts kept a tally of all the hacks and foals, but not of the packhorses:

Mr Roberts had just bought two or three new packhorses and one of them was a big chestnut mare – lovely-looking horse, but she's big, solid. And Mrs Roberts come down to the stockyards there one day and she said, 'Oh, that's one of the new packhorses. Come on now, you boys, which one do you call Winnie?' No one said anything and she said, 'I know you call the biggest and fattest horse after me!' And she had a good laugh with us – she had a great sense of humour, Mrs Roberts. A terrific woman I reckon.

In the early days when Lou Campbell was mustering at Ngamatea there were several packhorses that dated back to Ruddenklaus' time, before the Fernies took over. The musterers were right out the back 'as far as we could go', overlooking Taupo from the Dowden and they lost a packhorse there. 'It was a very narrow ridge and we had two packhorses tied to one another, and one got off balance and went down hill and pulled the other one. Both went down, and one was killed. We didn't know how we were going to put it to Jack Roberts when we came in, and all he said was, "You got enough sheep to pay for it." There were just the four of us mustering at the time.' In much more recent times the boys lost Winnie on the way to the Boyd. 'The poor old girl went over a bank, with her load. Wedd or Withers was sent back to take the shoes off Winnie and he was nearly in tears doing it.'[23]

> One time we were going to the Log Cabin, down across the Taruarau, up the big face and through the bush, with the horses carrying the big pack-boxes. One of them elected to take a shortcut through the trees; they were always vying for position in the pack line. It was the big part-Clydesdale mare called Kate. She comes out the other side of the bush – no gear on. She'd gone between two narrow trees which picked the boxes up, must have smashed the girth and the lot tipped off back in the bush. We had to take her back and fathom out the gear and tie it together and continue to the Log Cabin.[24]

Trooper was one of the great characters. He was a big horse, close on seventeen hands 'and as soon as you put a pack on him he'd try to roll it off or he'd go and scrape it off'. He used to carry the vegetables 'and if he got half a chance he'd bloody lie down and roll on them and mash everything before it was cooked'. He was loaded with chaff one day, a sack or two on each side, and he got down and tried to roll them off and of course rolled flat on his back and could not get up. 'We had a helluva job to get the gear off him and get him up.' The deer cullers had used Trooper: he was around the camp all the time with them and he had learnt all the tricks of the trade.

> At the Rock Camp there was a tent and a fly – you slept in the tent and cooked and ate in the fly – and they put the pack-boxes that had the bread in right in the back of the fly, which was just about in the tent. Overnight Trooper came in and lifted the lid on the pack-box and got into the bread and ate it. And nobody woke up! He could open the sliding rails but we were a bit cunning for him with the rails – we used to tie them so he couldn't slip them.
> There was one very tall horse, and he used to lean over to take the load then straighten up again. He was very clever at balancing. One person could load him – he was part of the loading up process so the packman-cook could put the load on one side at a time rather than loading both sides at once.[25]

Trooper rolling on his load; Sonny Barrett, Don Hammond and Young Tom Harker.

Opinion was divided over Cobber. He was either 'the quietest of the packhorses': he carried Morrie Mott off the Manson with his broken leg; 'a cracker of a horse': he carried a great big boar home whole for Dave Withers and Dave Wedd; or else he was 'a rotten old beggar' because he knew how to open the slip-rails and let all the horses go.

> We'd finished the muster, and old Cobber had opened the slip-rails on the horse-paddock fence at Golden Hills and led all our hacks and the packhorses back to the station by the Woolwash. The boys decided they'd walk in and pick up their hacks; some carried their saddles and some didn't and then we had to come back and pick up all the gear. You had to make sure you wired up the slip-rails otherwise he'd get them in his teeth and slide them loose, and all the horses would vanish. We used to take a tent with us to Kaimanawa for the chaff and a lot of our gear because there wasn't room in the hut for eight blokes and all the gear and one day out there we watched that Cobber at work. The old beggar was lying on his side and he had his head and neck under the side of the tent and he pulled out a bag of chaff, ripped it open, and he and two or three of his mates started into the chaff. We watched this happen.[26]

For all that, Cobber was still 'a great horse'. They always needed firewood and it was a popular pastime when they had a day off in bad weather teaching the horses to snig logs in to the hut. Cobber was good at it.

The stables at Ngamatea had a separate stall for each musterer and one morning Ian Sinclair was saddling up Gunner. Don McLean had the next stall. 'Gunner wasn't known as a bad horse, but he was a bit of a dirty brute and suddenly there was a commotion and Sinclair came out over the top of the stall and the horse left the heel marks of his shoes on the wall. I was in the stable when it happened. I saw him come over the top – and those walls were pretty high – and I saw the heel marks that were left.'[27] In later years Gunner was sent down to Pohokura for the resident shepherd but he was not an easy horse to handle and in the end he became a packhorse, and a good one. He found his niche at last.

Tarzan was an infamous character: he was the one they wanted to put geli and the dets on in the hope he would blow himself up. Although ththey didn't go through with it, he did come to a bad end, as Dave Withers remembered.

> The pack team was never allowed to operate with only one man. There always had to be two – one at the front and one at the back. There was a real art in being the one in front – he had to keep the team together, because they were never tied head to tail and sometimes there were ten or twelve – sixteen or eighteen in the full team.
>
> This day I was delegated to bring up the tail and Hal, our packman, was in the lead. We were moving from Golden Hills through to the Boyd and our mainstay packhorse was Tarzan, a big bay with a white blaze, bit of Clydie in him, sure-footed thing. He was at the back. You had to be careful they didn't try to put themselves in the front of the mob all the time, and start racing. We were going down a long face into Gold Creek and they got a bit of a run on, because Hal, in front, roamed on a bit too fast and let a bit of space develop and I got nearly to Gold Creek and here's this lovely big gelding Tarzan standing there with a swinging shoulder. The big horses carry the pack-boxes with the food in – about 100 pounds a side. With that weight he smashed his shoulder running down that slope. Nothing we could do about it. We had the rifle – just had to strip the gear from him and put him down. Very sad, but one of those things. That's why we were briefed by the head shepherd always to keep that pack team together.[28]

After the war there was a surplus of ammunition, greatly increased deer numbers and plenty of adventurous young men happy to be paid to take to the hills, so government deer cullers became a fixture on Ngamatea. They were allowed three cartridges per deer and were paid on their tally, and they had the option of skinning deer or taking lesser payment for a tail and a strip of skin up the back. They had a permanent camp down by the Woolwash and their packman took stores from there out to the cullers working in pairs on various parts of the station and packed skins back in.

Jack Bindon has had a long association with Ngamatea: he was a government culler at the end of 1949, then went on to be packman for the cullers for one season, and packman-cook for the musterers for two seasons. 'It was a great experience at Ngamatea. It was hard to leave there. I've retained a connection all these years. We went there on our honeymoon, and we took our first son back there and I've got a photo of Mrs Roberts holding him. If I go anywhere near I like to go into Ngamatea.'[29] Jack Roberts remembered the first time he saw Jack Bindon.

> We'd been out to Peter's on our ponies, might have been with Dad, and we were coming back in to the station and there was this group of about four or five new recruits who'd taken on government deer culling and the boss bloke was taking them out to Peter's. They had big packs and everything, straight out of town, walking out there, and it's quite a drag from the creek up the ridge towards Peter's hut and they were absolutely stuffed. I remember every little creek they came to they were having a drink to get the strength to get to the hut. That was the first time I met Jack Bindon – he was one of the recruits. Jack was a great big chap, and later on after he left Ngamatea the story goes someone bet him a dozen of beer he couldn't learn to fly, and he said, 'Right, you're on.' And he learned to fly and he was that good a pilot he became a flying instructor, and years later he showed up at Ngamatea in a little two-seater French plane, I think a Raleigh, a beautiful little thing. He'd flown all over the Manson in and out of all the gullies there having a look at his old shooting haunts.

Harry Bimler was also a packman for the government, then packman-cook on Ngamatea for a season, and he too retained his connection with the station. He left school at fifteen to become a deer culler and was at it for seven years. It was the beginning of a life of hunting, both in New Zealand and overseas. 'I packed to all the back-country huts and I shot on Golden Hills, the Boyd, the Manson, then down south and also in twelve other countries all round the world. And I've been breeding deer since 1961. I was New Zealand's first deer farmer – I got the first Rural Bank loan to go deer farming, in Rotorua.'[30]

In the 1960s there were still plenty of deer on Ngamatea to attract hunters but only those well known to the family were welcome to shoot up there, and some real characters who came to be known as the Three Musketeers were given what they would call 'exclusive rights' to shoot in

Two of the Three Musketeers: George Appleby and Sid Drinkrow, with Pam Forsyth and Hazel Riseborough; an afternoon shoot October 1965.

the Tikitiki bush. The original musketeers were Sid Drinkrow and George Appleby, who worked for the Ministry of Works in Napier as draughtsmen and bridge builders, and Rex Orange, who had a television sales and service business. Sid had had a long connection with the Fernies and with Lawrence. His father worked at Mangatapiri for 30 or 40 years, and he helped Gordon Mattson to get his job at Ngamatea deep in the depths of the Depression. Sid would spend his school holidays at Mangatapiri, staying in his father's whare and doing odd jobs for Lawrence, for very generous pocket money. He knew all the old-timers and saw three Fernie brothers together one time: Drummond, David and Walter – 'little skinny fellows, looking like swaggies'. He knew Rosie Macdonald too. After she left Kuripapango and lived between Fernhill and Napier she had a few sheep and she told Sid to choose one and kill it for Christmas. 'This one, Rose?' 'Yes, that one, Sid', and it's just kicking its last and she says, 'No, no, Sid. That's the wrong one.' 'Bloody hell, Rose, waddaya want me to do, stick its head back on?'

When Sid mentioned to his mate George Appleby that he had good friends at Ngamatea, the reaction was immediate – 'That's the place where all the deer are! Get in touch with them.' So Sid called Winnie, who talked to Lawrence, then said, 'Yes, you can come up – yes, all three of you – the Three Musketeers.' The first time they went to the station they thought they would have to go out the back to find deer but they were just unpacking their gear when Winnie drove in in the Land Rover: she had been out for the mail and had seen a mob of 24 in one of the paddocks on her way in. 'We tore out and got two or three! The rest just took off, over 8-wire fences.' This was the beginning of their love affair with the station: they shot on Ngamatea and at Pohokura for the next fifteen or so years. Occasionally Sid and George took their wives – their cooks! – Lee and Nita. Once when they took them in to Pohokura, they got there by about two o'clock and by four o'clock they had seen 204 deer. Other times they took them to the Tikitiki – they could get almost as far as the hut in their 1931 Model A Ford.

> What a marvellous machine – it never ever got a puncture the whole time we had it there, however long that was. The air was fairly low in the tyres and we had rope wound round them for traction. They'd put V8 wheels on it – they weren't the original wheels – it had 16-inch wheels on it. We bought it for £15, and we parked it up there, round the back of the storeroom at the homestead. That car could take any number of people – two sitting on the roof, two on the bonnet, five inside and a couple standing on the running board. We didn't have papers for it – we were about the fifth illegal owner. You didn't worry about that. We didn't bother registering it of course – it lived up there. We got very attached to that model A and we left it there in the end and gave it to Jack.[31]

They did a lot of the planning for their shooting trips in the Napier Cosmopolitan Club on a Thursday or Friday night, deciding when they would go, what they should take. They would be away for three or four days at a time, and one night Rex made a profound statement: they took far too much with them, too much weight to carry. The others asked what he thought they should leave behind. 'Well,' said Rex, 'salt for instance. It's damn silly for everyone to take salt.' And in the next breath he asked eagerly, 'How many flagons will we take?'

> Rex liked seeing everybody having a good time. He liked his beer. Sid and I liked it too – and we were prepared to work at it. We were only newcomers to it, of course. We were at Pohokura hut and we'd all gone different directions. That was the night Rex got lost and Sid had to go back for him. I got back to the hut a bit earlier – but it was dark, and I thought I'd put it across Rex so I opened a flagon and had a glass of beer. There were a lot of empty flagons outside so I got one, rinsed it out, and slopped a bit of beer into it. Sid had found Rex and brought him back and I put on an act that I'd been into this beer. Rex saw this and I came in with an armful of wood, and there was a little step there so on purpose I tripped over it and dropped all the wood across the kitchen floor. And Rex said quietly to Sid, 'He's half pissed – come on, we'll

fill him up.'³²

Later, when Rex got busy with his TV business, Geoff Keller took his place on the team and sometimes their boss, Roger Preston, a design engineer with the Ministry of Works, joined them. They gave him orders while they had the chance. Sid told him he must never shoot unless he could get a head or a heart shot: he must not waste good meat. So poor Roger would stand there for ten or fifteen minutes with a stag in his sights, waiting for it to turn on just the right angle before he dared shoot. He had to do as he was told; they were his passport to the place.

After Lawrence died they confined their shooting to Pohokura and got into pig hunting down there, sometimes with Denis Sheehan, another of their draughtsmen. Joe Campbell, who worked for Jack, was down there so they would go off two in one direction, two in the other. George and Denis were coming through the bush and they could hear a pig in front of them, a sow with a litter. They let her go, then sighted another.

> I put the musket up and bang – and he ran and disappeared and I thought, I've lost him. Had a hunt around and couldn't find him and when I told Sid he asked how big it was. 'Oh, a good porker,' I said. The next day Denis and I were back in the same area and I came on this little clearing where I'd had the shot from, and I thought: Now I shot from here . . . over there . . . and I followed the track – and I find this hole – and in the bottom a little pig. He'd fallen in and he'd been trying to get out all night, and he'd have died if he'd stayed there. So I pulled him out and Denis took a photo – this porker that I'd said he was – he was still a sucking pig.

'We loved Ngamatea, the silence, the nature. The bellbirds up the back of the Tikitik were beautiful, and they were just as thick singing in the scrub down in the Pohokura Valley. A real chorus.' Sid agreed: 'It was magnificent down at Pohokura – just like a big park. Beautiful, beautiful country, and we had some great times. But really I was more infatuated with the Titkitik bush, the wild background and everything. Those peaks rearing up round you.' And they were lucky: they were able to give something back to the place that had given them so much. George designed three bridges for Pohokura – the big one with a span of about 80 feet – and negotiated contracts with Eastern Industries in Hawke's Bay, who fabricated the bridges, carted them up and erected them. 'They went in after I retired. It made a difference to that place, cut out that Spiral.'

There were good cops and bad cops, and Lou Dolman was a good cop. He mustered at Ngamatea from September 1943 until April 1945, and was a brother-in-law to Harold Quilter, who had been packman for a season before the war and came back in October 1944 as head musterer. Lou was a great rider – the only one who could manage to stay on a little mare called Tornado. One time the boys decided Lou should ride Tornado because there was snow on the ground: snow builds up under horses' hooves and they thought this would slow her down a bit. 'He got on her and she got into it, and he stuck to her, but her feet built up with the snow and she lost her footing and went down. When she got up his foot was still in the stirrup and I remember Gordon Mattson, quick as lightning, raced in with a knife and cut his stirrup leather, otherwise she would have dragged him.' Later Lou joined the police and was based at Tuai. He became a well-known and revered figure and a big name in Search and Rescue in the Urewera: he received two bravery awards and an MBE. One day he saw Lawrence and Winnie at the Hawke's Bay Show and came up for a chat, looking smart and official in his uniform. Winnie said, 'The kids are in the tent there watching something – go in and see them.' Jack and Margaret were standing side by side when suddenly a great big hand descended on their shoulders. 'We looked up and here was this big burly policeman. We got a real fright. What have we done? Then I recognised him.'[33]

The era of good cops lasted into the 1960s, through the days when it was reasonable to think anyone trying to evade the law would probably hole up on Ngamatea and the police could just ring up and get a steer on their fugitives' movements. In turn they looked after the boys in town and helped deal with tricky station business like dead bodies. By the time Ray Birdsall took over as manager things were deteriorating; by the time he left it was like the wild west up there. This was the era of deer poaching for live capture or venison and there were police who were into deer farming and more interested in their property than Ngamatea's, and police who were into deer poaching and landed up in court. There had always been illegal shooters on Ngamatea, 'people coming in uninvited and unannounced. If they were to ring up, it was no problem, but if they went in unannounced it was a problem. You never knew where they were, and after all a phone call is a cheap admission price,' as Joe Larrington put it. There was illegal fishing too, especially after the advent of helicopters, although flying trout flies are less dangerous than flying bullets.

The better the venison prices, the more poachers there were. They left tyre tracks, footprints in the snow, and the station staff would track them, chase them and hope to catch them. Ray was a great tracker. He and Larry Cummings were heading out to White's block one day when Ray noticed motorbike tracks, which they followed towards the scrub and bush.

We found it cached up in a patch of scrub, down by the White's hut. We found sleeping bags, two of them, food, but no sign of the shooters, so Ray said, 'Go and get the tractor and we'll drive over it.' I said '*You* go and get the tractor and *you* drive over it.' He said, 'Ah, I think you're right.' So we picked the motorbike up and took it back to the station and locked it up in the petrol shed. Under the law in those days you really couldn't do much about poachers. The law said they could just come and pick it up. They went to the cops and said the station had taken it – and then they came and got it. And the following week we had the killing house loaded with deer – there was no chiller there then – and it was nothing to take 15 or 20 deer into town if everybody had had a good weekend. These guys had been back during the night and chopped it all into a thousand pieces. That was the station's payment for taking the motorbike.

Another outfit, and they never ever caught them, were walking in and floating the deer down the Rangitikei River on tyres and bits and pieces. They would have a pick-up under the bridge at such and such a time on such and such a day. You just couldn't stop them. [34]

Sometimes the intruders were after other things than deer. There must be easier stations to rob than Ngamatea, so far off the road, but the distance did not deter everyone. In the middle of one night the boys woke and saw lights on up in the workshop. Ali Robertson, the head shepherd, jumped into his ute and raced up just as the thieves leapt into their vehicle and took off. He chased them out to the road and halfway down the Taruarau Hill before he gave up. Ray and the boys had a 5 a.m. start one morning and Maxie woke up later to find that someone had been in the house. He had come in by a sliding door into the back porch and the laundry, found the gun rack and helped himself to Ray's semi-automatic rifle and ammunition. Maxie had just got a call from Chris Page at the cookhouse to say a fellow had come in and told her the manager's wife had said he was to get petrol. She went to the bowser and gave him petrol, but her suspicions were aroused by the immaculate inside of the car – not a tissue, a bag, a scrap of paper. Maxie called the police, who put up road blocks, but the thief had disappeared. It turned out that he had seen Ray and the boys riding out as he came in and had given them a wave, so they thought he was a legitimate visitor.

The police called Ray one day and told him to go out and block the Napier–Taihape road. Someone had been killed down in the Bay and they thought the killer was heading inland – that was the usual escape route. They knew he was armed and they instructed Ray to take his loaded gun with him and to 'shoot him if he goes you'. Ray drove up to the road, shaking his head in disbelief, and waited at the top of the Taruarau Hill until he asked himself what on earth he was doing there, and drove back to the station. Next day he got a ring from the police. They had found their quarry – dead. He had never left the Bay, and had turned his gun on himself.

There was a knock on the door one night and a cheeky poacher told the boys his vehicle was stuck on the Tikitiki track and he wanted them to pull him out. They told him he would have to go up and see the manager. Ray pulled him out – and took him to court for trespass.[35] One night they saw footprints in the snow and followed them next day out to the Tikitik where they found a couple of tents and guys still asleep. Ray made them break camp, took their rifles and escorted them back to the house and called the police. Two more prosecuted for trespass.[36]

> We had some weird situations with deer poaching. There was one chap who deliberately set fire to the tussock out there by the Tit. It raced across the Swamp, burnt the sheep, raged all day. There were all sorts of people there, trying to get the stock out. I remember Jack in his Land Rover, tearing through the flames. We didn't lose any cows and calves – they survived, but we lost all the sheep. He just deliberately burned the tussock to have us on. And we were continually having fencing cut.[37]

The police had no money for two-way radios, but Bert, Ray's friendly Taihape cop, wangled a couple from the MPs at Waiouru so the Taihape police and Ray could each have one. Ray never got his: it was purloined by the chief cop at Taihape who had his own deer farm and was expecting to be raided the same night by the same gang Ray had been warned about.

> He said, 'You'll have to look after yourself, we've got no spare men.' I went up to the crossroads there and waited till about two in the morning for them. By then I was getting too tired and I went home to bed. If they come, they come. I was woken just at daylight – I heard a vehicle coming down to the gate by the plantation. It was a Toyota – I watched it head over towards the buildings and grabbed my semi-automatic, jumped into the Nissan Patrol and there they were having a go at the bowser. Three jokers, all had guns, so I quickly loaded my gun. If you've got three jokers facing you all with guns, what are you going to do? I wasn't going to back down. I leapt out of the car and roared, *'Put those guns down'* and they quietly put their guns down. They knew I had a round up the spout. I took the number of the vehicle, the numbers of their guns, and then I had no option but just to order them out. I said, 'Out you go. I'm following you.' I didn't have the right to take their vehicle, too many of them to try taking their guns, and too vicious. They drove out, I

followed them at a distance. I'd told them any nonsense and you're going to get it – but I was by myself. They went right out and turned onto the Taihape road and headed off, and I tore back to the station and called the police. 'Right – we'll have road blocks out straight away.' They blocked the Taihape road, the Hawke's Bay road, the one to Pukeokahu and the other one. Those plates didn't belong to any registered vehicle, nor the guns – they weren't registered. I don't know who they were and they couldn't find them on the road, so they never ever found them. Probably headed into our Otupae block or somewhere, waited a day and then took off.

This was well on and by that time I think the marijuana was on the job. I wondered if that's what they were after, they had planted it out the back, or they were going to plant it – I can't remember the time of year it was. I never ever found any. But I was worried, very vulnerable, and all the Taihape police thought about was their own deer farm.[38]

The Ngamatea deer farm was just being set up around the time Ray left. They were using live capture. Ray had tried it himself but did not like it – the flying was too risky.

There was a terrible lot of poaching going on. They built airstrips everywhere – fixed-wing planes taking the deer out. Nick Koroneff had lost his land to the Army,[39] and he had a plane and he dropped in one day and said a parachute had just been dropped on the little airstrip by Peters, against the Rangitikei. We were working in the yards and we raced out there on our horses and found the parachute and the stores – bacon bought the day before from a dairy in Taupo – all dropped on the airstrip. They had shooters out there and they'd dug square holes about 6ft deep and the legs of the deer – just hocks, were piled up, ready to be flown out. We pinched their ammunition and all their tucker, we spread their rice over the airstrip and threw their bread in the river!

A truck appeared on the Otupae block with cases of ammunition and all this possum poison. It was illegal not to have it under lock and key but it sat there several days, with a chopper buzzing around. The police from Taupo called and said this joker was there poaching and had drums of fuel under his chopper. 'Put a bullet through it,' they said! I said to my truck driver, 'We'll go out and get that truck of theirs and haul it onto the back of our big truck and we'll lock it in our stables.' We did that, and later a plane came in – this crook from Taupo and his local police, and they'd picked up the police in Taihape on their way. He demanded his truck and I asked him for the ownership papers. 'Oh . . . er . . . didn't bring them with me.' 'Then get in the plane and go.' The police knew I was within my rights. He came back two or three days later with the ownership papers and I gave him his truck – and the police did nothing about him. You'd wonder what was going on.

Things suddenly improved when the old head of the Taihape police got moved on. Unbeknownst to Ray, complaints had been laid about this cop's attitude to Ngamatea and Ray in particular and the new head, a really top cop, came up to Ngamatea 'and quizzed me about poachers and all sorts of things. Then younger fellows came to Taihape from the fraud squad in Auckland – a complete change. We got to know the new head, a good fellow.'

A tractor disappeared overnight from up on the highest point of the station road. It was there when Ray and Maxie drove in quite late coming back from the Hawke's Bay Show, but it was gone next morning and every gate on the station road was lying flat, strainers snapped off. A bunch of thugs had come in in the night, got a vehicle stuck on the way out to the Tikitiki and hauled it out with the tractor. It was too much trouble to open gates so they rammed the strainers and drove over them. At least this lot ended up in court. The new breed of cops in Taihape saw to that.

The days of rampant poaching came to an end when it was no longer economic to capture deer from the wild or try to market the venison. Once deer farming became well established the focus was on genetic improvement through selective breeding and there was not the demand for wild capture. Then tighter regulations made it impossible to buy straight from the helicopters. Deer have to be quarantined before sale or slaughter unless they come off a TB-free farm so the market for feral venison all but dried up and the helicopters were put out of business. Poachers are still around, of course, but they are no longer a threat to life or limb.

Chapter Nine

Margaret

Ngamatea was Margaret's real love in life.

Margaret missed her mate Jack when he went away to school but the next three years alone at home, learning by correspondence, only increased her independence and self-sufficiency. She had a small menagerie of livestock to keep her busy, but her pony was her first love. She rode with her parents or D'Arcy or charged off alone, making the most of her freedom once the school day was over.

Joe Larrington remembered his first visit to Ngamatea when Margaret was about ten. He had gone up for the weekend with his brother Bill, a keen tramper who knew the country well; the Larringtons had come from England on the same ship as the Cravens and the two families lived near each other, then kept in touch over the years. As Joe and Bill were leaving Margaret decided to ride out part of the way with them. 'My brother was driving and I can still see her on her pony, this blonde hair streaming out behind her in the wind. It's a picture I'll never forget.' Two or three years later Lawrence needed a builder and he called on Joe. He and his wife Rita and their two small children 'put in a winter or two' up there. They lived in the shearers' quarters, and once when the cook took off Lawrence asked Rita 'if she would do a bit of cooking' for the boys who were there over the winter. They had happy memories of their time on the station and it was a change for the musterers, and for Margaret, having a family and young children around.

> I had an old Commer van in those days and as in all motors there were frost plugs. Margaret and the wife and kids and I were going somewhere and we were getting pretty near the road and all of a sudden the cab filled up with steam and I knew what had happened. One of the frost plugs had blown out. What are we going to do now? We had to find a source of water, and fortunately there was a creek down below us but then we had to find something to carry it in. Well, there happened to be Margaret's hat – the only container there was. Then we had to find a frost-plug. I had some coins in my pocket and there was a penny there. I managed to make it slightly concave so

it would fit in, and I jammed this penny in and gave it a tap with the hammer to make it sit right, then we filled it up with water with Margaret's hat. She rather enjoyed the experience. I don't know how many trips we had to make to the creek to get enough water – and only a hatful at a time. But it worked – and the penny stayed in there for ages![1]

Margaret had some bantams she kept in a run at the back of the washhouse. Further out was the fowl house where a dozen pheasant eggs, contributed by the Acclimatisation Society, were hatching under a hen.

> When they hatched they just took off, bolted. The hen frantically tried to round them up, but they were gone. Out of the whole lot one chick stayed behind and that was a cock pheasant and Mum used to have several roosters she was fattening for the pot, and every morning this cock pheasant used to stride up there and thrash these roosters, one by one, for his morning constitutional. And if he was feeling really fit he'd have the big rooster on, the boss. And if that wasn't enough he used to come down and give Margaret's bantams a tune-up too. He was a beggar, and we finished up – Mum and Missy and me carrying this blasted bird and we took him away down the Tractor Ridge and into White's and let him go – and never saw him again.[2]

Jack remembered his mother and Margaret doing a scenic flight from Taupo over the mountains. It was before Margaret went to high school, and they had 'young Seth Barnes and a chap Kay Rayner' with them. It got a bit bumpy over the mountains and the two men in the party started to feel pretty queasy. 'By the time they landed they were only too pleased to get back onto terra firma and Margaret looked at them, green as hell and not feeling very good, and she stuck her nose in the air and said, "And we women are supposed to be the so-called weaker sex".'

Margaret was prone to madcap adventures. She was eleven or twelve when the hydro scheme went in and there was the penstock in place, but not joined up at either end. She decided she needed to explore this thing so she poked her little dog into the pipe and crawled through behind it, down, down into the dark. All she could hear was the tick, tick, tick of the dog's feet on the pipe. It was black as pitch but luckily she and dog eventually popped out the lower end. Not a soul in the world would have known where she was.

When Margaret finished her primary school education she went to board at Iona in Havelock North and Morrie Mott or Ivan Chapman, two of the smaller and lighter musterers, would ride her pony – or ponies – 'just around the station, to keep them in trim'. But the minute she was home she would be out riding. 'She was a good girl. She always liked her

Margaret with Begorra at the Hawke's Bay show. Winnie Roberts is in the background.

horses. We'd see her around in the holidays – she'd turn up all over the place. She was a real tomboy I reckon.'³ She'd often bring a friend home from school, someone equally keen on riding – usually Belinda Wilson, sometimes Kerry McRae or Pauline Hildebrant. One day Margaret was riding her pony, Begorra, and Pauline was on Jack's old pony, Oracle. Bruce Atchison remembered the incident, with gales of mirth.

> They'd been out riding, past the haybarn, out that way, and on their way back they'd been having a bit of a canter. They got to the gateway and they're coming through and the pony Margaret's friend was riding, the older one, he went down. We were working at the cattle yards and Margaret said to her friend, 'Now you sit on his head so he doesn't struggle and I'll go down and get Dad.' So Lawrence sent one or two fellows up there and there's this girl sitting on the horse's head holding it down, but the horse has had a heart attack and it's stone bloody dead!

Academically Margaret had lost nothing by doing correspondence 'but she wasn't happy at Iona. She was used to the wide open spaces, and she did it hard having to conform.'⁴ Neither Margaret nor her friend Belinda was enamoured of boarding school. They met in 1954, their first year at school. 'We were both lost souls there, we felt isolated, controlled, but we had common interests and we soon teamed up. I was appalled that Margaret had never been to school. I'd never known anyone who had done correspondence and I really felt for Margaret living in such isolation!'5 For free spirits like theirs boarding school was very restrictive. Only on Sunday afternoons did they have any time to themselves, and on very occasional Sundays their parents could come and take them out. Kerry McRae's parents could seldom come down from Wairoa, so the Roberts would take her with them. 'We had some lovely days together. We just had part of the day so we could only go to Haumoana, or to the Wellwoods – they always had sumptuous lunches! And sometimes to Ted English's place, and we'd ride Ted's horses.'6

Margaret did a professional course at school, but was also able to study art. 'She was a very good artist – she did School Cert. art, but it was a real bind, very hard work. She also wrote well – I remember her picking holes in *Sunrise on the Hills* and she complained about it till I told her to write the story herself. But she didn't.'7 She could have done; as Kerry said, 'she had a gift for words – she always had wonderful stories of Ngamatea – stories about the shepherds, what her father had done. Ngamatea was Margaret's real love in life – and my brother, Lindsay McRae, was up there at the time, so I revelled in her stories. Margaret won the English and Oratory cup at Iona in the fifth or sixth form, and that was a top award.'

Occasionally Margaret went to Wairoa to stay with Kerry and Kerry stayed at Ngamatea a time or two. The first time she went they stopped at Rosie Macdonald's for a cup of tea. Rosie had a Khaki Campbell drake for Winnie, and Old Jack, Rosie's right-hand man, was instructed to get it. He put it in a sugar bag with its head sticking out and it sat in the back of the Super Snipe between the two girls. When they got to the Strip it was snowing 'so Mr Roberts ordered Margaret and me and the Khaki Campbell drake out and we sat under a tussock while he manoeuvred the car up the hill – in reverse. That was my introduction to Ngamatea.' Then she was offered Lawrence's grey mare to ride and all went well until it started to rain.

> We had parkas with us and I had mine half on and the mare started to buck. I managed to get the parka off and dropped it – and Margaret said casually, 'Oh yes, I forgot to tell you it bucked.'
>
> The Air Force used to practise low flying over the station and Mr Roberts had the mare hooked up to the back gate one day and the jets came screaming over and she took fright and pulled back. He was so annoyed he rang the air commander or whoever at Ohakea and told him off to a standstill, and they said – and this is genuine – there was this big notice on the board at Ohakea: 'If the grey mare's tied up to the back gate at Ngamatea DON'T fly low'!8

From the first year at school Belinda spent at least half of all her holidays at Ngamatea. Occasionally Margaret would go to Belinda's home in Levin for a few days, but then they would both race off to Ngamatea. Directed or encouraged by her mother, Margaret soon had Belinda helping with her household tasks. 'She taught me so much – it was all new to me! And you know, I called Margaret's mother Mum, like she did. Mrs Roberts was too formal, distant, so very quickly it got to Mum. But I didn't say Dad to Mr Roberts. Nothing probably.'

The two girls roamed far and wide on their horses – within reason, within reach; they couldn't stay out overnight. Lawrence took them further afield, to the Tikitiki and Otupae huts, but not as far as Golden Hills. Forty years later, when Belinda brought her husband Steve Dakin to Ngamatea, she recorded some of her memories of those days.

> I remember the endless red tussock as far as the eye can see, and remember upon instruction from L.H.R. riding from Timahanga to Ngamatea and back before lunch to fetch a whetstone (much needed) to sharpen the blades for dagging. I also remember riding from Timahanga to Pohokura for Leo's breakfast which consisted of a mounded plateful . . . two each of sausages, chops, bacon, eggs, tomatoes (wonder of wonders), and winching buckets of water from the spring behind the hut . . .[9]

Bruce Atchison reckoned he grew up with Margaret and Jack. 'There wasn't such an age gap. When I started they were about ten and twelve and I was sixteen.' Bruce was at Ngamatea on and off over a period of almost ten years from 1950 and stayed over for several winters.

> I was up at the homestead one morning, doing some job. Lawrence wasn't there – he was away and Margaret was home from school. They had a tomcat up there that was causing trouble and Marg's got the window open and she's got the shotgun pointed out the window. But the window was too high – she couldn't get the gun against her shoulder and the butt of the thing was half stuck under her chin waiting for this darn tomcat to come out and she was going to blow it to pieces. I said, 'Maggie, if you pull the trigger, with that kick you'll blow your bloody jaw off.' I can't remember whether she got the cat or not![10]

Margaret had about fifteen months at home before she went to Massey and that was the year an Australian girl, Margot Vincent, was at Ngamatea doing the books. She was lively company and they became good mates. When Margaret started the wool course at Massey in March 1959 girls were a bit of a rarity. Boe Gregory-Hunt from Pitt Island had done the first year in 1957 but then went home and married and stayed there. Then came Margaret, and she was lucky – there was another girl doing the

Margaret with pet merino lamb.

course with her: Mihi-Mere Lewis. 'I'm known as Missy,' she said. 'So am I,' said Margaret. They were a great pair, the two Missys, one dark-haired, one blonde, and both enchanted with wool. They only had three months at Massey, then twelve months' practical, and another three months at Massey. We arranged a placement for them in Auckland for their practical, and Margaret was determined to get me to Ngamatea before she went. I nearly did not make it. I had broken a leg skiing and torn the ligaments badly, but I persuaded the orthopaedic surgeon to allow me a week off from physiotherapy. 'Yes, you can go to your friend's farm,' he said, 'as long as you only walk on firm flat ground.' We did not enlighten him as to the extent of firm flat ground at Ngamatea.

Although we could not roam far that week, we drove down to visit Seth and Rose Barnes at Pukeokahu. We were coming back later than we should have and Margaret decided we would go in on the road, not the Strip. In no time the Land Rover was stuck fast. She commanded me not to move and climbed out into the mud to try to position the chains so she could reverse onto them. Nothing doing. She tried again while I got behind the wheel. It was really getting late and we were not going anywhere. 'I'll walk in to the station – you just sit there.' I was aghast, but she assured me airily it would not take long: she would take the short cut. I insisted on trying again and this time the Land Rover got a grip. 'Keep going, keep going. Don't stop. Left hand down a bit. Keep going – a bit further and you'll be able to turn.' I could not see a thing in the blackness behind me, but luckily the bank of the trench-that-was-the-road was almost non-existent for a short stretch and with Missy yelling instructions I pulled up with the back wheels on the tussock, shaking. She gave a whoop and came tearing after me. I slid over into the passenger seat – and sat on the great gobs of mud that had flown off the chains as she flung them into the back of the Land Rover. We laughed all the way home and burst into the kitchen. Winnie – Mrs Roberts as she was to me then – was standing at the Aga, stirring a pot. She looked at us, glanced up at the clock, and said calmly, 'I was going to give you another ten minutes then call the boys to go out and find you. Get into the bath, you're covered in mud.'

When Margaret finished her wool course she worked as a wool sorter at Tucker's scour at Clive. 'They thought a lot of her and I think she enjoyed working there. And it pleased her dad, having her working in wool. She did well at it. She would have done well at whatever she turned her hand to.'[11] When she married Terry Apatu at the end of 1962 Jack came back from the South Island for the wedding – he was a groomsman, and Alan Bond was the chauffeur. Alan had been kicked in the head by a horse not long before the wedding and lost half his teeth and had a broken jaw. 'But I had Uncle Walter's big black Humber Super Snipe for the weekend in Hawke's Bay, and a dinner suit on, and I thought I was a real smartie. I was always there, around the homestead, all the comings and goings. Just got treated as part of the family. Margaret would be home. We all loved her, of course. She was pretty outgoing – held no punches.'[12] When Margaret lived in North Auckland after her marriage it was a long way from home. She got back occasionally, especially after her dad became sick in 1965, but the next year she was in Waipukurau, not too far away from Lawrence and Winnie in Hastings. Lawrence died just six months after they got back.

After some years Terry came to have a close involvement with Walter and Eleanor Fernie and he got on well with them. They were an odd mixture of parsimony and generosity. They had their favourite charities, one of which was the YMCA in Wanganui where the Fernie Lounge was opened in 1974.[13] Terry would go down to Otamoa and stay for a night or two and bring them to Hawke's Bay for the show or take them to Wanganui for business and shopping.

> No one got on with Walter because they were scared of him. But if I saw something wrong, I said so. They lived in the past. Uncle Walter didn't

Winnie Roberts at the Hawke's Bay Show.

understand dollars and kilometres, Aunt Eleanor was very house-proud – she kept the curtains closed against the sun, and dust-sheets over the furniture. But it was she who got me involved with the family business. She had a financial interest in Mounganui; it wasn't making money and she called me in. That's where I started – supervising Mounganui and when the accountant and solicitor saw what I was doing they said I'd better look at all the properties.

Down at Otamoa Uncle Walter would want to come out and take a hand in everything. I'd get the hacks in in the morning and one would be saddle-sore, one wouldn't have shoes. Uncle Walter had plenty of shoes – a line of them at the back door. One morning he was putting odd boots on and I said to him, 'Uncle Walter, you're putting odd boots on.' He gave me a long hard look and said, 'Boy, I've got odd feet'.[14]

In the years when Winnie Roberts lived alone in or near Hastings she had lots of visitors, lots of reminders of the old days. The move to Te Awanga from a flat in town came after Margaret and Terry took her out

'I used to see the musterers bringing the black cattle down the hill towards the cattle yards. All those black beasts . . . sometimes there'd be four of the grey horses, prancing and dancing around.'

there 'to look at a cottage'. She asked if they were thinking of buying it for themselves. No, Jack and Margaret were buying it for her. She loved her new house from the start: the carpet was the colour of the tussock, where her heart had always been. She used to say that driving up towards Ngamatea, at a certain point there was the tussock, and she felt at home. There were many things Winnie loved about Ngamatea.

> I used to see the musterers bringing the black cattle down the hill towards the cattle yards. All those black beasts, and the hacks they had – well, sometimes there'd be four of the grey horses, all prancing and dancing around. We fed our horses with chaff and oats and they were always on their toes. They were positively dancing, knowing they were getting to the end of their beat and getting into the yards. It was a pretty picture. It was really. And it was nice too when there were thousands and thousands of sheep in the yards. That looked good.

She remained active, still driving her little Honda Civic up and down to Waipukurau, but one day, unexpectedly, when she was staying with Margaret, she had a stroke. Luckily the hospital was just minutes away but there was little they could do except nurse her with tender loving care and keep her comfortable. Margaret sat beside her for weeks, and came away each time and 'cried buckets'. Winnie slipped away at the end of February 1980. Her funeral was on 29 February, the sort of date that would have tickled her fancy. All the old hands from years back were there to say goodbye. Her ashes, like Lawrence's, were scattered by plane over the land they were part of.

❦

Ngamatea belonged to Margaret but she had not had a home there for years. She had always wanted a house in the tussock, and at last in October 1980 earthworks were started for the house John Scott had designed for her. He had sat up there in the tussock for hours and days on end, communing with the landscape, absorbing the magic of Ngamatea, feeling its healing quality, the way its encircles and enfolds you. It is vast, but it is not a repelling vastness, not an emptiness. There is something about it that is inclusive, that embraces you. The house that resulted is Ngamatea writ small: it encircles and enfolds you; the spaces are huge, yet intimate. John recognised Ngamatea House as the best he had ever designed – and his fellow architects recognised its uniqueness. On 19 March 1991 a member of the jury for the National Award for Architecture described it as 'One of New Zealand's special buildings, an amalgam of client and architect relating to the land and the most human of needs, shelter, warmth . . . materials played to the highest levels, it is simple, honest and belongs. Thank you John Scott.' The long process of building the house had been rather fraught. John was marvellously vague and not easy to deal or work with: he had no idea of time or money and drove the builders crazy, but Margaret could work with him, and appreciate his vision. D'Arcy, with his usual scepticism, was less enthusiastic.

> When Margaret was building that flipping house they couldn't get a cook for the builders and she said would I like to cook for them. They were camped down in the house near the quarters – there were only three or four of them. So I landed that job for a month or two, staying down at the cottage. A lot of the sub-contractors wouldn't stay – they came up from Hastings every day, so you can imagine what that cost, paid by the hour to run up and down that road.
>
> Then old John Scott would come up and the builders would come home ranting and raving. They'd finish something he suggested they do, they never had a set plan of the whole place, he just played it by ear, and he'd come and have a look and say, 'I don't think that's very good. We'll alter that.' He'd

Ngamatea House 'snuggles back into the landscape and looks as though it grew there'.

draw a picture in the sand and say, 'We'll do it like this.' They'd come home ropable. They'd have to undo all the work they'd done and start again. I remember they put a bit of wooden framing around one of his diagrams-in-the-sand and put a sheet of iron or something over it to keep the weather off so they could refer to it as they worked!

When the house was finished Margaret told John of her plans for the fence and garden, and he said, 'If you do that I'll come up and rip the fence out myself. I have not designed this house to have a garden.' Margaret planted flaxes and hebes and other small natives among the tussocks and the breeze ripples through it all and the house snuggles back into the landscape and looks as though it grew there. The first time the family went up to stay the curtains had not been hung, and it was snowing. They built up a big blaze in the great stone fireplace and put the mattresses down around it. Once the place was finished they made only one change: 'We put in a freestanding fire in the nook by the kitchen so the family could gather there instead of in front of the big fire leaving Margaret alone in the kitchen'.[15]

In an undated, unsourced magazine story written by Iain Morrison, probably in early 1983, we get Margaret's view of things.

> 'Ngamatea is a very personal matter', she says. 'People never seem to get the perspective right. I was born there, grew up there, and watched my parents

Ren Apatu on the front terrace of Ngamatea House.

struggle with every problem there is. The only incentive scheme in their day was for roading, and my parents faced cartage charges, low wool prices (their only income), staff shortages, staff problems, isolation, and lack of money. . . . I guard jealously this heritage and intend keeping it intact for our children. After we rehoused all staff we built ourselves a home on Ngamatea. We don't enjoy travelling to and from Waipukurau. The house will give us a base.'

Running a property like Ngamatea from Waipukurau has not been an easy task, and Mrs Apatu is quick to praise the two managers who have run the farm since her tenure began. Ray Birdsall . . . . 'is like a member of our family . . . . he knew the situation on Ngamatea. He is a tremendously good manager, and he set things in motion again with very little money.' . . . When Ray Birdsall left Ngamatea Phil Mahoney was appointed manager. 'Again we were lucky. Phil came at a time when the farm was in good shape and with the government schemes in operation he was able to carry on the development which Ray Birdsall started.'

Ngamatea House and the station under snow.

In November 1984 there was a grand reunion at Ngamatea for the opening of the new eighteen-stand woolshed. There were over 300 people at the Saturday night buffet dinner and most of them slept the night on the station. Over 30 of the guests were warming Margaret's new house. They put up stretchers in the garages and put down mattresses on the mezzanine; some clever souls put their sleeping bags on the lambskin-covered window seats in the nook and others filled all spaces in between. Maxie and Ray had pride of place in a big guest room in the east wing. Down at the station the quarters were full – and the cookhouse, the manager's house and the cottage, and there was a tent and caravan city set up nearby.

D'Arcy, Elaine Mahoney and Sheila Maxwell brought greenery from Boyd's bush to decorate the shed and there was a fishing net full of balloons strung from the rafters. Bill Jones had brought a big gang from Fernhill to organise food, music, entertainment, campsites – and put down a splendid hāngi. A huge effort had gone into preparing the food – sandwiches and salads for lunches, Sheila's 20 pavlovas and a *huge* fruit salad for dessert. The buffet was set out in the woolroom, and the shearing board became the dance floor. Before midnight the dancers moved back and Margaret took the microphone and thanked everyone for all they had done for her parents and for Ngamatea over the years. Then she introduced Bill Jones, whose son whipped a sheep out onto the board and shore it, and the shed was well and truly open. With the formalities over the balloons were

Bill Jones's boys putting down the hāngi; the cookhouse is in background.

released and the band played until four in the morning for those who could last the distance.

Earlier that Saturday Margaret had crammed her four-wheel-drive Toyota van full of people and led a tiki tour out to Owhaoko trig and beyond. Many of the boys had not been back for years, had not seen the transformation from tussock to grass. They were astounded to be able to drive in minutes over country they had only ever ridden over, slowly. Frank Brady expressed their thoughts:

> When I was at Ngamatea I couldn't see anything else but the farming methods they were using. I was absolutely amazed when I went back and saw what Margaret and them had done with the ploughing and the grass that it was growing, because, really – it was all pumice, and if someone had said, 'You're going to grow clover and real grass on this stuff', I'd have laughed at them. It's amazing what the super and that did. They poured it on – and when I went up for that woolshed opening there was 8 inches of clover growing everywhere on this barren land that we used to muster. There were places there where you went for miles and there was just pumice laying on the ground – and you thought how the hell would that ever grow grass. It was just barren land, that's all it was.

Paul Thomas and Dave Withers drove out together behind Margaret's van, most of the way to the Tikitiki bush then up onto the trig, but when

Margaret speaking at the opening of the new woolshed; Bill Jones on the right.

Margaret turned back towards the station they went the other way, out beyond the Swamp. It was only a year after the fire, and the firebreaks made driveable roads.

> It used to take us so long to ride out there, it was all tussock and sort of little wet guts and creeks and bits and pieces, and you could bog a horse in there so easily. And now you could just drive over grassland. There are a couple of lovely old rocks there and you look down on the Lake block, and the Pinnacles are just sort of up behind you. From there on hadn't been developed so you had all the tussock and we were right back to where we were 24 or 25 years before the opening of the woolshed. We sat there, facing the Pinnacles, with our backs up against the rocks and I'd brought out a chilly bin with numerous cans of beer. The sun was going down just to the left of the Pinnacles and this lovely warm breeze came through the tussock and the snowgrass, just sort of whispering away and it was absolute bliss. The sun went down and it started to get darker, but it was still warm as toast – this lovely breeze and this whispering of the tussock. And then the silhouette of the Pinnacles highlighted as the sun went down. All that and a bit of beer inside you! Eventually we got into the truck and decided we'd better get back to the woolshed because although we'd mustered that country and knew it so well, as it was we were getting a bit lost in all this grassland and we thought if

The 18-stand shed in action: eight stands on the left hand side.

we don't get cracking before it's totally dark we'll be in trouble. And when we got back we were greeted with, 'Where the hell have you been? We were just about to send out a search party for you.'¹⁶

Horses were a big part of Margaret's life, and hunting became more important to her once she had more time, and her base at Ngamatea. Her visitors' book recorded the visits of endless hunt clubs who went up to hunt each season, and old Ngamatea hands would come across her hunting, especially round the Taupo region; she was always glad to see them and have a yarn. After a while she bought a property at Hatuma for her horses – Ngamatea horses, just her hunters for a start – but she soon had young ones she had broken in that needed plenty of work and she was lucky to find twin sisters, Wendy and Robyn Jacobs, to give her a hand. They were working on their parents' dairy farm in the Bay of Islands; Wendy was not long back from Seoul, where she had been grooming for the 1988 New Zealand Olympic team. The girls came down for an interview early in the new year. 'It was that absolutely terrible drought and the further we drove the worse it was. We never thought New Zealand could get like that. We got quieter and quieter, and we got to the stage we were so horrified neither of us was talking.' When they hit Waipukurau Robyn said, 'It'll have to be a bloody good job for us to want to move to here.' But once they met

Margaret they decided she was 'a really neat lady' and they never regretted for an instant their decision to move down and work for her. They were back in a couple of weeks with five horses 'in a little old truck', and not much else. They moved into the empty cottage at Hatuma and the first week Margaret gave them a bonus and told them to go and buy themselves some furniture. 'And we did, and started work – and we were there for seven years – seven good years.'[17]

After about three weeks of drought-stricken Hawke's Bay Margaret said, 'Let's go to Ngamatea.' They stayed six weeks and 'absolutely loved it'.

> But that road frightened the hell out of us. We got halfway up and we were both sick and it ended up with Robyn driving, me sitting in the passenger seat and Marg sitting in the back. That's how we used to have to go to Ngamatea, the three of us, because Robyn and I could not handle the road for years. The scenery was beautiful but – oh, we couldn't appreciate it.
>
> We stayed up there a long time. We brought the horses in and Marg gave us both a hunter – Robyn and I had never really hunted seriously, and at that stage she had the horse float that carried three horses and we went everywhere from Ngamatea. We went over to Taranaki hunting, we went to Rotorua, to Taupo, or she'd say, 'I think we'll go down to the Manawatu today – it's only a couple of hours.' And off we'd go. We met a lot of her hunting people and Dannevirke hunt, where she was a member, would be one of the last hunts and we did all of them before we moved back down to Hawke's Bay.

They went back to Ngamatea every year for up to a month – between the Horse of the Year and the new hunting season when the Manawatu or Hawke's Bay Hunt Club would come up to join them. They would bring their cooks and fill up the house and the quarters – and all the hares on the place would be put to flight. The ten or a dozen Ngamatea brood mares Margaret started with were running down on Mounganui with their foals and they would be weaned and taken down to Hatuma.

> Marg was just getting started. Then we had a few good sales – a couple of Ngamatea-bred horses that we schooled at Hatuma, that sold well and she got really keen on the breeding. We got a new stallion with good blood lines, then she brought down more mares and we got bigger and bigger so Marg employed Oliver Edgecumbe too. He was a really talented rider and he learned to break in and the three of us produced them and took them round the shows and sold them.

They soon had so many horses that they needed better transport and Margaret bought a new truck – just cabin and chassis – and Brian Hobson built a horse-box onto it, with space for five horses and living accommodation as well: Margaret's home-away-from home. The girls updated their vehicle and could now carry seven horses, and with Oliver along they would take them all to a show – and ride them all; these Ngamatea-bred horses were becoming well known. It was hard but they loved the life, loved working for Margaret. Like her father, she knew her staff: she knew that when the girls were having trouble with a horse it was time to back off and give them some room.

> She was a really neat person. Sometimes she'd disappear and then she'd come back and say, 'You guys are running late and I don't feel like cooking. I'll go into town and get us some tea.' And when she worked with you she wasn't your boss, she was one of us.
>
> She did some hilarious things. I remember one day taking off from that little set of yards, just down from the house at Ngamatea, where we used to keep our horses. We'd had terrific snow, and Marg decided we were going down to the station to watch the crutching and instead of going back and onto the track, we just hurtled off in that big blue machine of hers down that hill, in the snow, and it got into a sideways slide – and it just slithered. Robyn and I were gripping the sides, petrified, and Margaret is driving along, and she says, 'I hope this stops soon – we're going to hit the old road in a minute and we'll be on our side.' And we slithered right across and ended up parking on the edge of it. I don't know what didn't move us that extra foot. And somehow we tottered off again in a very low gear, and we got away. And she was unfazed by it all, absolutely unfazed.
>
> And the first time she ever drove the Hobson – we used to drive it 'cause she didn't have her heavy licence, for years! – she took a group of ladies up to Auckland to see the élite international riders competing, and they camped on the showgrounds in the truck so they could be part of the action. When she came back Robyn asked how she got on driving the truck and she said, 'Really good, really good – but I hit a bank.' And Robyn said, 'Oh, yes, that's a terrible road out there to Isola, a terrible road.' And Marg says, 'Oh, not that sort of bank – the ASB in Ponsonby. I parked, and I didn't allow for the camber, and the truck tipped.' When she went to drive off it ran straight along sort of in the gutter and the top of the truck tore the front off the bank![18]

Then there was the time the three of them caught a poacher. They had ridden out around the Tikitiki bush and suddenly there was a little pup tent, and two hinds hanging in a tree behind it. Somewhat taken aback, Margaret pulled up and sat looking at it for a minute, then she said, 'Come on girls, we'll ride over there.' They stopped short of the tent and Margaret called out to see if there was anyone at home.

Out shot this guy and he was in such a panic he was just in his longjohns. He stood there and Marg started into him: 'Who are you? What are you doing here? I am the owner of Ngamatea.' It ended up he was a policeman and Marg gave him a real towelling down. Eventually it stopped and he bustled back into the tent, terribly embarrassed and apologetic. As we rode away Marg turns round and she says to Robyn and me, 'Now – lesson to be learned, girls. You've always got the advantage when you've caught a man in his underwear!' There were heaps of funny things she did but that was typical Marg.

There was a change in management at Ngamatea at the end of 1990 when Phil Mahoney left. Margaret and Terry offered the job to Graham Lunt, who had managed Smedley for 20 years. They knew Graham; their son Renata, had been a Smedley cadet in 1987–8. Graham had been a Smedley cadet himself, then a single shepherd there, and after his marriage he went back as assistant manager with his wife, Margaret, then managed another property for five or six years before returning to Smedley. He was there for 28 years altogether.

> The idea initially was we'd go to Ngamatea for five years and as it turned out we stayed exactly five years. It was a challenge. I would have liked to go there earlier and be there longer but five years was quite a good stretch at that stage of things. Terry would come up about once a month, but he was always going on to Mounganui anyway. If we were selling stock or anything I always used to tell Margaret and more often than not she'd come up.[19]

After the years of development Graham felt it was time to consolidate. There were tracks, gates and gateways that needed attention, paddocks that needed regrassing and especially subdividing. They were cropping about 240 hectares in swedes and chou and the odd crop of greenfeed oats, grazed with a hot wire, and they put in a lot more shelter. 'They had planted quite a lot of trees, and we carried on with it and put in about 5,000 pines a year, the years we were there. The idea was to get timber out of it but you'd no sooner planted than it would snow and all your work was gone in a night.' They were years of big snow. It would lie centimetres deep for days at a time 'but when it fell most times it only seemed to fall on one-third of the property, so we had two-thirds clear'. They were snowed in time after time, and if they could get as far as the gate they would not be able to get down the Taruarau Hill. They had got snow at Smedley, but not every year, and not like this. 'I got to the point where I wouldn't lamb ewes up on the front country. It was all so exposed so they all came closer to home.' They even lost the roof over the covered yards in October 1992: it collapsed under the weight of the snow.

We ran cattle out as far as they'd go – around Hawkins, across Poverty Creek and out to the back of the Lake block; it was an outlet for about 1,000 cows. They'd winter out there and come back in prior to calving, then in summer after weaning we'd put ewes out there for a couple of months, and they went all around Hawkins and Wild Dog Spur and all around there. We were never over the Ngamatea boundary. That post with Ngamatea on, on the back boundary was there. I think Ren put it in. They didn't go far – but they certainly spread out – all up around Peter's. But then the Conservation fellows came out and they reckoned stock were eating the place out. They fenced off a couple of areas out towards Burglar's, to see what the difference was between grazing and non-grazing. There was none in my opinion.

Marty Crafar was still head shepherd when Graham started, but after a year or so he went down to Mounganui as manager. There were four other shepherds, and Glenda and Arthur Lee were cook and tractor driver. Sheila and Gordon Maxwell had had to move down country in mid-1987 when relations with management got sticky, but as soon as their jobs came vacant again Margaret went looking for them. She saw Gordon in the street in Waipukurau, where he and Sheila were both working, and offered them their old jobs back. As a precautionary measure Gordon said he would like to have a chat with the new manager before he decided. They went up, had lunch with the Lunts and moved back at the end of March 1993. 'It felt like we'd been overseas for five years and come back home.'[20]

Graham and Margaret Lunt had been at school with Maxie Birdsall in Waipukurau, and later Margaret nursed with Maxie. Her training came in useful when the shearers' cook had a heart attack.

> She'd actually stopped breathing. I'd rushed down, and someone got onto Rex Selwyn, who was just there in the yard with his truck. He was one of the super-spreader fellows and he'd just finished a CPR course, one of the requirements for his part-time job with the County. I had my finger on her pulse and he was doing mouth-to-mouth. We kept her going till the chopper arrived – from Palmerston! It would have been quicker to come from the Bay, and they took her to the Bay because that's where she was from. Everyone was calling her Aunty and before she went in the helicopter I said, 'What's her name? – Let's write it all down.' I thought, If she dies in the helicopter, whose aunty is she? But she survived.
>
> You never knew what you might have to do up there. Of course I was always feeding somebody. I used to go to Hastings once a month and buy the stores for the cookhouse. I'd drive down by myself, and go to Rattrays, wasn't it? Used to go round the shelves, with trolleys. I had a storeroom – you should have seen that. It's still going now – down at Sheila's. We had the storeroom

Stags in White's block.

up at the house – just opposite the back door. Not the old room, from the old days. When we left they built an extra storeroom for Sheila, and they made a little office for Gordon.

When Graham went to Ngamatea he was surprised at the number of people coming onto the property to hunt deer – and all acting as though they had the right to be there. 'They were coming out of the woodwork, ringing up all the time, and quite curt too. Anybody, everybody. I didn't know where I was.' As far as he was concerned the staff had first option, and he curbed the influx of outsiders. 'I got a ring from a cop in Palmerston who said he was coming out to shoot – *said*, not asked. I think he was the same one Margaret caught in his underpants, out at the Tiki bush, and he ended up going to jail.' The cop who went to jail, whether or not it was that one, was a witness for his mate when he appeared in the district court in Taihape charged with illegal hunting and giving a false name and address.[21] The pair had been caught down in White's Valley where they had set up camp deep in the scrub. The defendant was fined $350 and costs and the matter may have ended there, but there was such concern over the 'quality of the evidence' given at the trial that the case was referred to the solicitor general. This resulted in the witness, the other member of the hunting party, being charged with having perjured himself, being convicted and gaoled and dismissed from the police force.

View from Ngamatea House towards the Pinnacles.

Margaret had another encounter with a cop when she ran her horse truck into a bit of a ditch 'around Hamilton somewhere'. It was not a major problem but this cop drove by and asked her, 'Where's the driver?' 'I'm the driver'. 'Who owns this truck?' 'I do.' He was obviously surprised. 'Who's this M. C. Apatu, Ngamatea and Hatuma, on the door? I go up to Ngamatea shooting every year.' 'Oh, do you?' said Margaret. 'Well, that's me, and I own the property. I've never had any contact with you.' 'That set him back on his skids,' said Graham, 'and he had to be very nice to her then!'

During the 1990s the SAS held annual exercises in the high country. It would be around early April and they'd base themselves at the Ngamatea shearers' quarters and park their helicopters in the horse paddock.

> They'd let these jokers go at the Taruarau, and they had to get back to Waiouru, across country. They had a team trying to catch these young jokers and stop them getting through, and we used to find them out in the paddocks when we were mustering or whatever. Margaret and I were out looking at a new fence line – Ren might have been there too – and one poor little joker came up and said did we have any tucker – he hadn't had any tucker for two days. Marg went home and got some stuff from the house and took it back out to him.
>
> Another time half a dozen of them stayed in the house overnight – they bedded them down and rubbed a bit of grease on their blisters. Marg was

enjoying it like anything. She'd fed some of them that time and she said, 'You come back to the house at dusk and knock on a certain door, so we'll know it's you.' These jokers were frightened the searchers would get on to them. But Ren got some old cartons and these jokers climbed into them and he took them down to Ohinewairua in his wagon. I think the Ohine boys used to take their dogs out in trailers and they put the dogs out and these fellows got into the trailers and they took them right out the back of Ohine to get them on with the job.[22]

Ren had enjoyed his stint at Smedley under Graham's tutelage: 'He was a stalwart traditionalist and he showed us how a traditional sheep and beef station should be run. We even learned how to load a packhorse, how to balance loads and tie them all up, the real basics – shear and crutch and run a dog. It was all sort of demo stuff, large station stuff writ small, but it was all there.' Then before he went on to Lincoln he had a summer on Ngamatea in 1988–9. 'My dogs were used to overshepherded, very docile Romney sheep at Smedley. I put a huntaway out around the Ngamatea fine-wools in Stone Quarry – a huge mob – 10,000 … 11,000 hoggets, and they took off and pelted down to the bottom of the hill where Matt Burke was waiting. He was also used to Romneys and they all descended on him. He got them through the gate alright – but only just!' Phil Mahoney was manager, John Roberts head shepherd and Ren just another shepherd, 'a junior one', living in the quarters, eating at the cookhouse. They were still riding horses, but there was a fleet of Toyota four-wheel drives and two-wheel Suzuki DRs – white and blue, 'just the flashest things'. They did a lot of shooting and took their Suzukis even up to Peter's. 'John Roberts – no relation, by the way – was a mad keen deer shooter. He was pretty good, but he got lumbered with everyone else who wanted to go. God knows how he shot anything with all us in tow, but he did.' The summer was 'a helluva lot of fun' – and not only that: it was good for him. 'In retrospect more than I knew at the time. I didn't actually realise how good it was to know the quarters and know your way around.'[23]

In subsequent summers Ren worked at Mounganui, on a Banks Peninsula dairy farm, and at the meat works at Takapau. His three-year degree at Lincoln, a Bachelor of Agricultural Commerce in farm management, required a fourth year of practical work. His Smedley years covered the cropping, and sheep and beef components, but he was still required to have fourteen weeks on a dairy farm – and that had to be in the South Island since the rest of his practical had been in the north.

With Lincoln behind him, Ren was back at Ngamatea for the 1992–3 summer, as tractor driver working with Arthur Lee and with Graham Lunt as his boss again. When Arthur and Glenda left early in the New Year Ren

Ren the tractor driver at Ngamatea, 1993; tube silage, Case tractor and Honda 200 bike.

had the job of picking up about 1,000 big round hay bales which were sitting in the paddock 'going unround – they'd sat out there far too long. Farmers' Transport used to back-load super and we'd take the truck for the day and load bales from the paddock and dump them at the hay barn and I'd stack them. It had all come to a bit of a grinding halt because Arthur had left.'

In the early 1990s John Scott designed another house for Margaret at Hatuma – very different from the Ngamatea house, but with his characteristic brilliant use of space. It is light and bright and airy with a Mediterranean touch. Again he sat on the hill and communed with the countryside and designed just the house she wanted for her age and stage. Her marriage had ended, and she was fully occupied with horses and Ngamatea, and building this new house. 'It took forever to build – they put stuff up and tore it down. Then it sat a while till other contractors turned up and finished it off.'[24]

The Lunts' years at Ngamatea were coming to an end just about the time the Jacobs twins found a piece of land down near Norsewood which they could afford, and where they could start their own horse training business. They hated to leave Margaret. She had recovered from surgery after a fall from her hunting pony, but then further health problems surfaced and they felt 'really terrible, like we were letting her down, walking away when we shouldn't have been. That last month working for her we were really divided – excited about our own property, but we were all really hurting. It was like the end of an era, like leaving home again. But she was good – she came in and saw us a few times that year, so that made it a lot easier, and we'd go up there and help her.'

The Hatuma cottage was free and the Lunts were going to be looking for a place in Waipukurau. 'Marg suggested we stay in the cottage and look after the farm. We lived out there about six months then bought this place in town.' Apart from the horses, Hatuma was used for finishing Ngamatea stock, especially surplus cattle, and managing a 180-hectare block was just the right sort of retirement project. Graham is like so many old Ngamatea hands: they cannot quite retire, they just move to smaller properties – anywhere, as long as they can keep their dogs. 'With foals and everything I had 90 horses. Margaret used to hop on the back of the motorbike and we'd go round and see all the horses. They'd all come up to the motorbike, and she'd pat this one and tell me the breeding of the next one. Once a week, or once every ten days or so. They were her great love.'

Graham's place at Ngamatea was taken briefly by Peter Watts, who had been his head shepherd. It was an awkward time to be at Ngamatea and it did not work out well for Peter and his wife Jenny. When they left Margaret appointed her head shepherd, Steve Kelleher, as the new manager. He had been at Ngamatea about nine months, living in the married shepherd's house with his wife Keryl. Steve had been a Smedley cadet in 1970–1 and then Graham asked him to stay on as a shepherd; he had worked on the Kelly block with Phil Mahoney, and for Frank Brady at Olrig for six years. He had heard a lot about Ngamatea before he ever got there.

> I've probably got the best farming job in New Zealand because I'm stock manager and all I really have to concentrate on is looking after the stock; I've got a married head shepherd and four others working for me. Ren pays the bills and grows the grass, Gordon Maxwell runs the machinery part of the deal and takes care of the agricultural work. All the bookwork and politics that Ren has to get involved in I just wouldn't be interested in and I can concentrate on the hands-on job and getting it done.
>
> I had managed a few small properties but I actually got away from farming for ten years, then I came back in – and for ever and a day I will be grateful to Margaret because she gave me a huge opportunity, something to sink my teeth into. She presented me with a challenge – and it's a biggie! One of the things

you learn at Ngamatea is nothing will be perfect and you can't stop trying for it to be perfect. I love this place with passion, but I tell you something – it keeps you humble.[25]

Margaret read her man well. Steve is a dog trial enthusiast and an excellent stockman, as well as a good team man. Ren, Steve and Paul Hughes constitute the management team Margaret put in place, and they work together well. Ren, who had had three and a half years as field officer with Ravensdown Fertiliser in North Canterbury after his Lincoln years, brought with him the experience of working with every type of farmer and farming enterprise when he came back at Christmas 1997. It was a perfect background, but he had not set out when he left school to equip himself for the job.

> Ngamatea was so remote in a way. We just visited. I didn't see myself working there, running the place. It didn't cross my mind, not till after Smedley, Lincoln, Ravensdown – three lucky breaks that all came together. I met heaps of amazing people: converting dry-land merino farms into irrigated dairy land; I went from the Corriedale Farmer of the Year Award to winter lettuce in one day. That's how diverse the whole thing was. A whole lot of great out-there doing-it people. Fantastic. You went there basically to sell fertiliser but mostly it was opening the door and having a good look round someone's farm. Good management – how to get hold of something and give it a good shake. And better than that – no one knew who I was. Yeah, I wasn't the son of Ngamatea. You had to stand on your own two feet and get on with it.

They were great years for Margaret: she was farming her own land at last. 'Ngamatea is amazing,' she wrote about this time, 'looking so good, and a credit to my team there. Sometimes I can't think why it took so long to come to grips with what I have, and stop depending on other people. There isn't anything one can't do oneself really, some of us just take longer to get to it!' She did not get it all right from the start, but when she got a great team around her they did wonders. Weaning her off wool was a hard task though. Early in Steve's tenure she took him to the South Island to buy halfbred rams.

> We'd flown down and it was all teed up, but when I got there all I wanted to do was say, 'No, no, Margaret, we're on the wrong track here.' They were wool-producing sheep, but they had no figures as to their fertility and that. For the direction we were heading, we were actually buying the wrong type of sheep. Well, instead of bringing ten home, we brought three or four. In hindsight, one of the slowest things to our progress has been that we didn't

change our genetics. We actually said way back that we needed to change our breed of sheep, but Margaret wouldn't move with that. She thought it was the halfbred that would survive here, but what will survive here is what you feed. If you feed it anything will survive, and that's where the deal's changed.

Five years ago we started doing 550 hectares of winter crop – kale and swedes, which has been absolutely huge – not only to getting through the winter, but the spin-off from that 550 hectares of new grass after the crop is tremendous. So what we've done really is pick the biggest ugliest paddocks and put a winter crop in, put it into new grass and then most of them have been split either two or three times, and those paddocks are carrying fifteen ewes per hectare plus most of them have got 50 or 60 cattle as well. Basically we've got a ten-year programme – we're about halfway – then we start again and regrass it again, and hopefully the next time round it will be a direct-drill job. But the potential of Ngamatea is absolutely huge – infinite's the word.

Early on, when Margaret gave me the opportunity to do what I'm doing and I felt I needed a bit of help, I called on Paul Hughes's advice quite regularly, and we started driving around together and forming plans. It eventuated from there really. He ended up being a huge contributor to Ngamatea.[26]

Margaret, too, turned more and more to Paul. He was still the station vet, but he was much more than that; he was a consultant.

I had a lot of time for Margaret. She could be stubborn, she wasn't always easy to get on with, but I saw a lot of her, and came to understand how she saw things, what Ngamatea meant to her. She wanted to know the place was in good hands and was looked after; that was more important than anything else to her and she gave me the task of seeing that it happened. And it is happening, as she wanted it – the place is going ahead in leaps and bounds. She had the solace of knowing her management team could do it. She was always keen to come with Steve and me if we were going for a drive around the place and what really pleased her was to see that the stock was well fed. And she saw we were fed too – she'd bring along a bag of good Hawke's Bay fruit for a healthy snack for us all as we drove around! We never thought of taking anything, of course.[27]

With snow always a possibility in springtime at Ngamatea, picking the best time for lambing was never easy. Over the years there would often be a dump of snow at a crucial time but in recent years lambing has been staggered to lessen the risk of losses.

The rams go out to the first half of the mixed-age ewes the last week in April; a week later to the second half; twelve days later to the two-tooths; and a

Ngamatea-bred ewes with 'composite' lambs.

month later to the hoggets – a fair proportion of them get in lamb. We use a composite ram – a Finn, Friesian, Dorset Down, Romney cross – and it's been worth gold. We can keep the progeny of those composites and put more composites over them – and there'll only be one sheep like it in New Zealand because Ngamatea's almost been a closed flock. Yes, our quality of wool will drop and our weight of wool will drop, but we'll go from a halfbred sheep giving us maybe 110 per cent lambing to 130 per cent plus.

At scanning time all the twinners go to a crop. All the singles might get fed silage or finish off a crop, so we can utilise the feed. Our break fencing is pretty much strip grazing – a paddock of 50 hectares will be cut up four or five times with electric fencing. We employ a couple of casuals to do that work and they are almost permanent rouseabouts – fixing fences, gateways, killing a cattle-beast that's broken its leg, dog tucker, whatever.

It's a far cry from the days when Ray Birdsall came and there was one fence, at the back of the Swamp. But I'm kind of . . . envious? – is that the word? I would have loved to have been here in those days. What we are now is a great big lamb factory. We still have horses – and occasionally they get ridden. That's quite sad really.[28]

The cattle are an Angus-Hereford cross, which has proved satisfactory. 'There's probably no reason to change that. They are pretty hardy girls that

Cows in tussock in the Road block.

can still go and live in the tussock for four months of the year – and that's where we do still use horses. We try very hard not to calve anything in the tussock nowadays, but there's still a couple of hundred end up calving out there.' The deer farm, set up in the early 1980s, still operates with about 600 hinds and deer replacements but the profit has gone out of deer farming so it does not have the prominence it once had.

Hay and silage making is contracted out; Daljeet Singh from Palmerston North has done it since 1996. He is part of the place, and a real character. The hay is made in round bales, each the equivalent of fourteen conventional bales, and the amount made each year depends on need and the season. The silage is cut from surplus, usually not until February given the late season up there, and with forward planning the stacks can be situated strategically next to crop paddocks and on a paddock about to be cropped. This facilitates feeding the silage once the crop is finished, and raises the fertility of the paddock on which it is fed out.

TB has been a problem in station cattle on and off since the early 1990s. It is not known for sure how TB came on to the property but possums and ferrets are both commonly implicated in TB transmission. In 1993 Larry Cummings came back to the station to do pest control, working on an hourly rate and with a vehicle allowance. 'But it was too big a job for that arrangement and when possum control really started to kick in a lot of it went contract, per hectare, and Martin Brenstrum and his team have done it since 1997.' Just to complicate things at Ngamatea, pest control is the responsibility of two regional councils – Horizons on the western side, and Hawke's Bay on the eastern – and Larry won a monitoring contract for Horizons. 'I've been trapping out there to establish residual catch rates. Martin has done a marvellous job – the possum population is virtually zero but ferrets and feral cats are still a problem. Ferrets were carrying TB, cats are everywhere but there was no TB found in them and the last time I was out there we hardly got anything – one possum and one cat out of eight lines on one block.'[29]

Martin had been an enthusiastic part-time possum trapper for years when Margaret asked him if he would like to give up school-teaching for full-time possum control. He started with a contract for 2,330 hectares of White's, Centre Ridge and the Tikitiki bush.

> Margaret absolutely loathed having TB on Ngamatea and I was determined that I wouldn't fail the 5 per cent performance target – but I didn't actually know how much input was needed to achieve it. After a few months it became obvious that I wouldn't finish the job by myself and I had to get some help from Dave Sharp, Andrew Christison and Adrian Moody. We killed around 4,000 possums and I was hugely relieved when the monitor result came in at 1.5 per cent.
>
> From 1997 to 2001 we killed 10,396 possums, then Animal Health Board funding for TB control was increased to cover most of the station and the next year we got 2,900 in the scrub and beech beyond the pastureland. But interestingly we hardly caught a sick-looking possum and the few sick ones we delivered to Agriquality proved not to be TB-infected. One of the keys to good possum control is access and we spent a lot of time cutting walking tracks and ATV tracks to improve the access through the scrub and beech forest. We've been able to establish a very effective long-term baiting strategy along our network of tracks.
>
> In 2000 we began targeting ferrets and the station has funded the bulk of that work. We've got over 300 ferrets and about 15 per cent have been found to have TB. There's more to be done there.[30]

Getting and keeping good staff at Ngamatea was a difficulty in the early days because of isolation, today because the bright lights are so attractive and so accessible. The farming industry is short of labour and there are complaints that no one will bother to hire and train young staff. Steve felt he should do his bit and employed two young fellows straight out of school.

> I'm not sure I'd do that again. It's too taxing on the whole system to have two young ones at the same time. In recent times I've employed four or five

Smedley boys; they at least have some grounding. What we lack mainly is experience – we run huge mobs of stock and every time you send someone away to move 5,000 ewes you live in hope that they're good enough stockmen to get the job done effectively. They used to stay a year. Now basically they're not allowed to come in the door unless they're going to be here eighteen months or longer. When I job interview them I tell them not to bother coming unless they're prepared to be here for at least a couple of seasons.

Perhaps I'm just getting old, but young people haven't got the same dedication to doing the job as they used to have. Times have changed and with staff shortages everywhere there's plenty of jobs on farms closer in. And the romance of the back country has gone. I still dream about what it might have been like to have been here 50 years ago, 40 years ago. And that's all it is, it's a dream.

Sheila and Gordon Maxwell have been at Ngamatea for years. 'It will be sad when I leave here,' said Sheila, 'but things have slowed down – fewer visitors and people.' 'And farming has changed so much,' added Gordon. 'No place, no job is perfect, but there'd be a thousand worse places to work, and I don't think there could be too many much better.' Margaret once asked Gordon if he had a brother: 'We could do with another one like you'. Sheila reigns supreme in the cookhouse, and these days ordering the stores is up to her.

Ren is very good. He says he wants the boys well fed, and as long as you keep things within reason it's fine. They do long hours, and they should be well fed. I'm not a restaurant cook, but they get a good meal – and pavlova once a week, puddings every night – and they like their roast. Mutton, beef and pork are station meat. We fatten four pigs, one for each house and two for the cookhouse. Mutton is killed on the place, but not pork and beef; they go down to Hatuma and the home-kill butcher-man is just round the corner. Our vegies, eggs, frozen things come from Taihape on a Friday. There's a good supermarket in Taihape now and when you're in town you can walk around the store and see what they've got, and ask them to get things you need. And of course the road's much better. It's only 18 kilometres now to the tarseal, and the storeroom here at the cookhouse and the office for Gordon and the new double garage have made this place.

There's no store at the homestead like the old days. The boys can't buy gear – they can go to town so easily. I do the mail twice a week, Monday and Friday – or Thursday if I'm off for a long weekend. They come past the mail-box every day and turn there and go back via Otupae and Mangaohane, but it's too far for us to go out for the mail every day. I keep busy in the garden and mowing all the lawns – along the driveway, round the single-men's

Tom Whimp's wagon in the Ngamatea cookhouse garden.

quarters, casual quarters. It's not a ride-on mower – one you walk behind. That's my exercise. I do it about once a week. I've always done it since I've been here, just to keep it tidy – and get me out of the cookhouse.

All our buildings are marvellous. Before they built their own house here all our living quarters were beautiful. All the facilities are totally up to scratch, and so many stations just aren't like that.

Margaret took an interest in everything that was going on on the place. The staff were always pleased to see her: she knew everyone and stopped to chat. Richard Whittington was back in the late 1990s with his tractor to give Gordon Maxwell a hand.

We were quite often working paddocks that I'd worked originally. That was quite neat, doing them again after 20 years. It's interesting to keep up with the place. Like the old musterers said, the tussock gets in your blood. Everyone you talk to it's the same. I don't know – whether it's the size of it or the tussock, I don't know what it is. It's just a special place.

And Margaret was . . . she was an excellent lady. When she came up she'd always stop and talk to you and see how things are going. Excellent. No plum in her mouth or anything like that . . . I mean – she was a top lady. When our son was born she brought flowers and everything in to Karen in the hospital – a little pewter mug and that, and well . . . gee . . . here I was, just a worker and I thought . . . man, this was something. She was a top lady.

Margaret would tear up and down the road to Ngamatea in a little four-wheel-drive Toyota truck she had at one part of it. 'A nightmare of a machine,' according to Ren, 'overpowered, short wheel-base, and she'd dive in her handbag looking for her cellphone and address book. She wouldn't look at the road, she'd look at the paddock, the cows. You know the story about her hitting the bank? Well, that sums it up.' She always had Biggles, her Jack Russell with her, and he owned the passenger seat.

> I remember I went up to Ngamatea one day and she came up behind me with dinner and she stopped to talk to Jack on the side of the road. She had one of those cooked chickens in the back of the truck and when she got there Biggles was as big as a barrel – so full of whatever he'd eaten he could hardly breathe. And Mum said, 'I did have a cooked chicken for dinner, but I stopped to talk to Jack and Biggles climbed in the back and ate it.' He'd eaten the whole bloody thing, except the wishbone. We nicknamed him Mr Chicken after that – he did have a thing for chicken. Of course Mum used to feed him whatever was in the fridge. She didn't buy a lot of dog food – he'd be dining out on cube roll, blue cheese . . . whatever was in the fridge Biggles got to eat.[31]

Margaret spent time with Steve, keeping in touch with whatever was happening on the property. He had great respect and admiration for her and what he liked best was driving her around and 'skiting as to where we were at'.

> Marg was always delighted to come back here in the early spring. She didn't particularly like coming in winter. Obviously this place had had some pretty hard winters and got bashed around, but seeing stock that was being fed took away some of her concern about coming.
>
> That big acreage of winter crop made the difference. Instead of feeding hoggets we feed ewes and fatten cattle. We've come from 25,000 odd ewes to 42,000, in seven winters. But that didn't happen by accident. It happened by Ren and Margaret and Paul Hughes and myself working out a way to get through the winter, because if you've got a period of 120-odd days' nil growth, and you can't get through with 25,000 ewes, how do you get through with 42,000? You feed them better.
>
> This property could carry 70,000 ewes, but I don't want to be here when it does! It's become a pretty intense, high-pressure job and at times I get totally stressed with it, but I think we've got it sussed. I didn't come in here as any glory seeker – but I'd hate to leave here and think I didn't make a difference. And you're always aware of all that's gone before and the input of all those other people and that you couldn't be where you are if it hadn't been for them.[32]

Christian Findlay, Beatrice Findlay, Margaret, and Joy Beamish at Ngamatea House, 24 May 2000, 'sitting on the patio, a glass in our hands admiring the view of the wide-spreading grassland'.

Margaret was never short of visitors at Ngamatea: all the old hands wanted to come back. The visitors' book recorded the visits of Dr Bathgate in January 1988, Morrie and Margaret Mott in February 1990, and Ray White and family the same month, Maxie and Ray Birdsall for a weekend in March 1991, Joe Studholme in May 1992, Gordon and Eunice Mattson in April 1994, Lizzie and Christian Findlay in December 1995 and Christian and Beatrice Findlay, his mother, in May 2000. 'Home again,' she wrote, 'what a pleasure.' Beatrice Findlay was Guy Shaw's daughter and this was her second visit to New Zealand to discover her roots: the previous time had been in 1975. She and her son were staying with Marion, the widow of her late cousin Pat Donnelly, and they all had a day at Ngamatea. Margaret took them tiki touring. It was an emotional homecoming for 86-year-old Beatrice.

> I can't tell you what a great feeling I had driving up that road to Ngamatea – Gentle Annie, the Blowhard, and the glistening snowclad mountain top of Ruapehu. The thrill of seeing the award-winning house and meeting Margaret. Sitting on the patio, a glass in our hands admiring the view of the wide-spreading grassland which was my old home. A little drive to the waterwork in the river that engulfed my father . . .
>
> Meeting the manager and the extension officer was a bonus, past groves of pines, up and over the grassy ups and downs for the next hour was stunning.

Margaret, Dave Wedd and Arthur McRae, Ngamatea House, November 1998.

The shearing shed, pens, action spot so neat and clean and shepherds' rooms, manager's house among pines. I should have asked Margaret to show me the old house site. I'm sure I would have recognised some signs. So great to find out that Margaret and I spent our early years in the same house.[33]

In late November 1998 a mob of old-timers arrived with their mates and their horses to do a nostalgic trek around their old haunts. Dave Wedd, Arthur McRae and Alan Bond drove down from Wairoa in a couple of trucks with another farmer friend, Eric Steed, and their blacksmith mate, Albert Horsfield. 'Margaret met us up there at the gateway. She was going through to Taupo with horses and she said, "Don't you leave till I get back." We didn't, we were there for about a week, and we had a bloody ball from then on.'[34] They rode out around the Tikitik, the Tit, out to the Boyd, and back again. Jack had Peter Looker staying and they came up from Timahanga and spent a couple of nights at Ngamatea with them. 'We had a great weekend – Peter and I drove out to the Tikitik with all the tucker and these others were riding and we finished up in the bush, in a clearing, having a big meal – Bondy was cooking for them. And it was snowing like . . . well, it was *really* snowing!'[35] 'Cold, man, it was cold – just freezing. We camped in the shearers' quarters. Margaret came back to her house, and we had a good old yarn about the old days. Margaret didn't do the ride – she wasn't fit enough, but she came down and had a feed with us in the shearers' quarters. Jack came up, and Steve Kelleher took us all around the new development too. Great to go back.'[36]

As Margaret's health deteriorated she remained positive and kept up her usual pace as long as she was able. She was her father's daughter – and her mother's: part Winnie's compassion, part Lawrence's steel. Ren had married at the end of 1999: Sally Baines was a neighbour at Hatuma and Margaret had known her for years. Arthur McRae expressed everyone's thoughts when he said, 'I'm pleased her son had taken over while she was still there – that made her day, seeing Ren settled down and taking over.' She would be up and down to Ngamatea, keeping her finger on the pulse, seeing to all the projects and improvements she wanted in place.

> Her big ambition was to have that loading ramp just in front of the woolshed. The old one had been there for ever – an old timber one. So we got it all organised and completed, but she didn't see it, which was a shame. Everything else she wanted done, was done. How she worked! She was strong, and it wasn't a case of being pig-headed or anything – she was just mentally strong, and physically strong. She was back and forth to Auckland for treatment, but she'd never mention it. She'd just get in her car at Napier airport and arrive up here, then drive round the farm with Steve or see what the boys were doing up in the yards – testing the bulls or something. It was like Ngamatea was her best therapy.[37]

She was still following the fortunes of her horses at shows and events: Kerry Chalmers saw her at a one-day event at Takapau. Sheila Maxwell remembered Christmas Eve was the last time she saw her, driving around the station; and the first weekend in January 2001 the Jacobs twins spent time with her at the Dannevirke Show. 'We had no idea she was as bad as she was. Every time we saw her she looked all right. She was battling away and she'd never tell you. You'd hear rumours of she's real brave, but she'd never talk to us about it, never complain.'

Ren and Sally were in the cottage at Hatuma and in mid-January, just before their son was born, they called the twins and invited them for dinner. Wendy and Robyn were excited at the thought of seeing Margaret and they had a great evening with a group of her old friends. 'She was a box of birds, just the same old Marg. We were all getting dozy and she was the last one still talking, still being the great hostess, just like she always used to be. And it was only another two or three weeks . . . and . . . she'd gone. We were . . . horrified . . . just . . . absolutely devastated.'[38]

Margaret's funeral reflected her life. Many of the old Ngamatea hands had died in the previous few years, but everyone who was left must have been there. It was a truly fitting farewell, with the church and reception

Loading out wool today: over 100 bales per load. Shearers' quarters and the cookhouse in the background.

area decorated with Ngamatea tussocks, carefully dug out, balled up in sacking and tied with plaited binder-twine. When Jack stood up to speak at the church everyone held their collective breath for him, but he managed brilliantly. He talked about their childhood, then reminisced about the time their parents were away for the day and he and Margaret were to go down to the cookhouse for a meal.

> Margaret was twelve or so – before high school anyway, and starting to notice the boys. She got all dressed up and got into Mum's scent – plastered it behind both ears and on her wrists and we went down for lunch with the boys. We were sitting there in the cookhouse and Bruce Atchison came in. He started to sniff – sniff here . . . sniff there . . . and then his nose pointed him in the direction of Margaret. She's sitting there all dressed up and feeling nice, and he looked at her and said, 'What the hell have you been rolling in, Margaret?'

It was just the light touch that was needed; but then he turned to the casket and said quietly, 'Goodbye Missy.' Her hunting truck, her favourite vehicle, was parked outside the church, ready to take her on her last journey. The casket was put on a bier of hay bales in the back, the hunting horn sounded and the truck drove away.

Home was the hunter, home from the hills.

Chapter Ten

The Ngamatea Family

The place had an aura all of its own, and you were part of it.

Interviewing musterers and other old Ngamatea hands threw up an interesting phenomenon: many of them had little idea what year they went to Ngamatea and no idea how long they were there. 'Must have been about twelve years,' they would say, and the wages register would show it was eight, or they would say eight years and it would turn out to be four or five. They were not lying; they were not even exaggerating. This was their reality. Ngamatea was a huge period in their life: it stretched over years . . . must have done to leave such an impression. They were young when they went there, there were no radios, no newspapers, they had no calendars – no watches in most cases, they were not far into their working life on the land and what they learned at Ngamatea and what they gained from it set them up for life. The days and weeks and seasons ran into each other, and they came away with a new appreciation of life and work, and with an attachment to Ngamatea that never left them.

They wanted the stories told, but not all of them were keen on the tape recorder. The very first interview – with Sid Drinkrow and George Appleby – was a disaster. There was a problem with the microphone and the tape did not record properly so the story was there but only in snatches. Every time there was a burst of laughter the tape cut in or out. Luckily I had taken a lot of notes, for there was no repeating the exercise. I had not seen the Musketeers for 20 or so years, but they were at Missy's funeral in February – and George Appleby had not been expected to live into the new year.

It was some time before the next interview, but there was Gordon Mattson at 91 with his sharp memory. He chattered away happily until the tape recorder was turned on, then stiffened up and spoke very formally. We jollied him along until he forgot about it, then he filled up four sides of tape and said, 'Come back tomorrow', and we filled another one. He had a list of names he had drawn up a dozen years before when he had not been well and needed a pastime: names of blacksmiths, truck drivers, deer cullers; names of blocks and huts and creeks and rocks, of mountains and rivers, of packhorses and station hacks. He could reminisce about all of them. I

Interview with Gordon Mattson, Bell Block, New Plymouth, 28 March 2003.

asked him the name of a certain rocky peak in a photograph. 'Oh, that's the Tit,' he said. 'Yes – that's its real name.' 'Oh, no,' said Tony Batley later. 'Its real name is Te Toka a Ruawhakatina – he was one of the ancestors.'

I found the next 90-year-old, Lou Campbell, and his wife Joan, on a small property near Taihape. When I said, 'I believe you're ninety, Lou,' he said promptly, 'No. Not until April.' I tried again: 'And I believe you're still blade-shearing your own merinos.' 'Only some of them', he admitted, 'just my pets.' Many of the 'Ngamatea family' wives had an input – they had heard all the stories too – and various of their offspring would find an urgent reason to call in that day because Dad or Grandad was being interviewed and they wanted to listen in. And all the old hands wanted to know about each other, who had been interviewed and who was going to be and did I have so-and-so's address. The list of informants Jack and Lance had started me off with grew, and grew.

After the 90-year-olds came the 80-year-olds. Ray White, at 86, was living on the coast north of Napier; many of them live out of town, as near to a rural setting as they can get. Ray had written his memoirs, by hand, pages and pages of them, but 'only for the family'. Luckily he had given Jack a copy of the Ngamatea chapters so I could pore over them later. Bill Cummings, a mere 77, had a house in Taihape and managed to spend half a day there so he could be interviewed, but he would rather have been out at the farm. He went just about every day 'to help his son out'. But, like the rest of them, once he started talking Ngamatea he decided he was not in such a hurry to get away after all.

Bruce Atchison was high on the list – everyone came up with Bruce's name. When I phoned he said yes, a morning or an afternoon interview, it didn't matter. But then he looked at the calendar. Oh, Wednesday, no that wouldn't do. Cattle sale Wednesday; he couldn't miss that. Bruce is one of the great Ngamatea characters and the stories poured out, in the most colourful language and accompanied by great rumbling, shaking laughter.

Ray Birdsall was an essential informant, of course. I had met him a few times, at Ngamatea and at his farm, but when I phoned he was very non-committal. 'Ngamatea?' he said vaguely, then 'I'm not a talker.' He finally agreed: 'Yeah, Thursday will do.' I arrived, expecting a difficult interview, but we settled ourselves comfortably, and Ray started to talk – and talk, and talk. His face lit up, his eyes were shining, he was back at Ngamatea in his youth. We filled up side after side of tape and as a story would run out he would sit back in his chair, then suddenly he would think of another one and lean forward again. Finally it was time to stop, and as I gathered up my gear he stood up and looked at his watch. 'Good God, it's ten to two!' We had sat there for four hours. He turned around and saw the sumptuous smoko Maxie had left under a throwover on the table. 'We didn't have our cup of tea.' He started to make it, then said suddenly, 'We should be having lunch,' but we were both just ready for smoko. It was as though time had stood still.

Both Jack and John Roberts had said it would be hard to get Noel Roberts talking, and it was: he was even more allergic to the microphone than Gordon Mattson. But when he thought the ordeal was over he started to open up. Luckily I had only slipped the pause button on, and when the tape rolled again he was unconcerned, and we all had a good laugh. Noel was still farming; he had tried retirement at Taupo for a few years, but could not settle to it. They had a lovely view and Eileen said one day, 'Look at the lake, Noel. How blue it is.' 'Humph,' said Noel, 'it's not green grass.'

Sometimes we were lucky and got two interviews at one sitting: Dixie McCarthy called Trevor Topliss to come and join in; John Ruddenklau invited Lindsay and Shirley McRae along; Dave Wedd was down from Wairoa and came to Jack and Jenny's flat in Hastings – and Jack and Dave bounced stories off each other all afternoon amid gales of mirth. Jack and Dave were at Napier Boys' High together, as boarders; their memories went back a long way. Gordon Grant would not hear of anyone driving out to Porangahau to interview him: he'd 'drive a million miles for Ngamatea' and it was no problem to come in to Ren and Sally's at Waipukurau. There was an overnight trip out Porangahau way later to interview Bev and Bob Ralph and see their video of Johnny Boyd's visit to his childhood home.

It was a long drive to Wairoa and Gisborne, but worth it. Dave Withers, like any number of others, is still farming – but way up the Ruakituri

Interview with Joe Studholme, Coldstream, 3 March 2004.

Valley, halfway to the Urewera, where he is a Search and Rescue man. Next day I went on to Alan Bond at Matawhero where he was mine host at the Jolly Stockman. When he left his fireside for work in the afternoon Arthur McRae sat down in his place. Arthur was in Gisborne for a few days jacking up his next cattle drive. 'I'm the last of the drovers I think. There's not much done now. Not many guys of our age can still ride a horse – and I don't like them too rough, either! I still do a bit around Wairoa, a couple of mobs, Tiniroto way. We've got some cattle grazing at Te Reinga, and there's a sale on tomorrow. I'll stay and have a yarn, then back to Wairoa.'

Frank Brady was another who had not quite managed to retire. When he told his boss he was retiring, but admitted he would be keeping his dogs and would do a few days here and there, the owner of Olrig told him to stay put and do his few days a week right there: 'This is your house, your garden. Stay.' Peter van Dongen had had to slow down after heart problems: he and Joy had just settled at Ngahinapouri on a dairy unit which their sons would run. D'Arcy Fernandez was going down to Timahanga from his home way up north; he would stop off at my place for an interview and check out the fishing while he was here. He had lost Nat early in 2002 and the following year he scattered her ashes up on the Tit: 'There was snow all over the ground. Poor old Nat – I can hear her moaning yet. But it's a great view from up there.' Eric Brooking had lost his wife too and Margaret had gone to her funeral in Dannevirke. But Eric was keeping well. 'Healthy

Ngamatea back boundary post: Lance Kennett, John Roberts, Gordon Maxwell, Sheila Maxwell, Rob Laidlaw, Hazel Riseborough, 9 February 2004.

place, Ngamatea. Never lost a day sick and same now.' Don Hammond had retired, but said he would not be available for an interview; he was going duck shooting, around Gisborne. But he got talking about Ngamatea, Winnie and Lawrence, the old days. Then he admitted he had just had a hip done, had not had the stitches out yet. 'Yeah – I doubt I'll be going duck-shooting. After talking to you I reckon I'll renege. Yeah, I'll be here.' He was, and his son Craig could not stay away. He drove down from Tutira to Hastings for dinner so he would not miss out.

When I interviewed Paul Thomas we remembered our first meeting 42 years earlier – in Perugia, Italy. Paul was touring with three other Kiwis; I was working there. We were surprised when we introduced ourselves. 'Paul Thomas? – from Ngamatea?' Paul was at Pohokura when Missy drove me down the Spiral and as far as the Taruarau on the broken-leg visit to Ngamatea. We did not dare ford the river, so we did not see Paul, but now here he was. The Ngamatea connection stretches far and wide.

Contacting all but one of the early lessee families was a highlight of the project. At the beginning it seemed an unlikely possibility, but as it turned out they had all visited Ngamatea and Timahanga in recent years. Jack and Jenny remembered Beatrice Findlay's visit and directed me to Marion Donnelly, who was still in touch with Beatrice's daughter Heather Woollacott in South Africa. Jack and Jenny knew the Studholmes too

Golden Hills hut, 9 February 2004.

and when I finally contacted them they were just about to leave for the North Island. They would meet Ren and me at Ngamatea and go on to Timahanga and Hastings the following day. 'Then you must both come to Coldstream and see the family papers, and the Lindauer portrait of Renata Kawepo.' It was a profoundly moving moment for Ren to encounter the lifelike portrait of his eponymous ancestor looking down at him from the wall of the Great Hall at Coldstream.

Elaine Chapman was as great a Ngamatea enthusiast as Ivan, and more than most she wanted the stories recorded.

> Margaret wanted all this written down. She asked the boys to write their memories down – and she wanted to build a big cairn at Ngamatea recording the names of all her boys. 'Bring the boys home,' she said. 'My boys.' She was really funny. It used to make me laugh with this 'my boys', and my boys were just about all older than her I think. That must have been after Morrie Mott's funeral, because I think that was the last time we were talking together. Yes, it was, so many had gone, and she'd been to all the funerals.

Having met so many of the people, it was time now to get closer to the land, to the back country I had never seen. I had flown into the Boyd a few years earlier and walked north to Poronui; my ambition now was to fly in again and walk south to Ngamatea. When I mentioned it to Lance Kennett,

Above Gold Creek, Boyd Rock on skyline, 9 February 2004.

he said, 'If you do arrange that trip, I want to come.' Ray Birdsall said the same, but in the end he could not make it, and when I said, 'Well, you'll be able to think of us out there', he replied simply, 'I think of Ngamatea every day.' Others were keen but finally it was agreed that it would be Lance, me and Rob Laidlaw, who had not worked on the property, but had connections to it. But then Sheila and Gordon Maxwell and ex-head shepherd John Roberts wanted to come, just for a day trip, and bring a quad bike to carry the packs Sheila was sure we were not fit enough to carry. So six of us set off from Ngamatea, with *two* quad bikes; some tramping trip! We dropped our gear at the Golden Hills hut, now in a state of disrepair, had a cuppa there and ate some of the copious food Sheila had brought, then tramped on up through the Golden Hills bush and along the almost non-existent packtracks down to a ridge overlooking Gold Creek and the Ngaruroro. Three of us spent the night in Golden Hills hut, kept company by the spirits of all musterers past, and next day we turned back, still with the hope of spending a night at Peter's. Low cloud put an end to that, but we did explore the Tiki bush and climb to the rocky outcrop at the top of the Tit, in a howling wind. A month later Sheila and Gordon climbed with me to the trig on Peter's and at last I felt I was getting the hang of the geography on the ground, as well as on the maps.

There had been a Ngamatea reunion at Crownthorpe in June 1976: grace, and the Queen, and a printed programme with photographs and

View from the top of Peter's; the Donnelly and the Dowden on the skyline.

a merino ram's head drawn by Harold Quilter. Margaret planned the reunion with Bill Jones, who had then been shearing at Ngamatea for almost 20 years. It had been a great occasion but now, in 2004, it was time for another reunion – this time a pre-1972 one. It was organised by Murray Taylor – Hurry Murray, they used to call him because he was always on the go – and it was held in July in Havelock North and at Ngamatea and Timahanga. About 90 people registered and some nearly did not make it: the Napier–Taihape road was closed by snow all Friday, and the Napier–Taupo road was closed for most of that day. But by one route or another they got there. Any number of those present had done only one or two seasons on the station; others came back after 50 or 60 years. They had never forgotten the place, never lost touch. It was testament to the land and to the people who farmed it, the atmosphere, the experience – the magic of Ngamatea.

Friday night was 'time to catch up with friends from the old days'; Saturday was the projected visit to the stations, weather or not. About nine in the morning it was announced that the road was passable and the visit was on. The snow did not prevent a drive out towards Owhaoko trig, but only the four-wheel drives could make it up the hill so they ran a shuttle service to get everyone up there. Unfortunately the view was not perfect: the cloud was down on the tops, so Ray White, the oldest of the old hands

Aerial view of the station today.

present, provided a commentary. Standing in the snow and stabbing his stick at the cloud, he indicated all the geographical points around the horizon. 'You can't see them', he said, 'but they're all there.' After a memorable 'light lunch' and visits to the woolshed and machinery shed the next stop was Timahanga and afternoon tea, followed by a drive down to Pohokura and a serious nostalgia session outside the old cookhouse there.

The reunion dinner was back in Havelock North on the Saturday night. It was only minimally more formal than the previous evening in that Peter Looker acted as MC, proposed the odd toast and told stories that brought the house down. Jack proposed a toast to absent family and absent friends: 'One thing Mum and Dad really enjoyed was seeing the people who'd worked up there come back bringing wives and families. It really appealed to them to see some larrikin they'd known in the past all wedded and settled down with a family.'

Max Mossman, who mustered at Ngamatea for just the one season in 1958–9 wrote to Jack after the reunion and talked of 'the legend of mustering at Ngamatea' and of the privilege it had been to work for Lawrence. 'He passed on to me skills with horses etc that made me a much better person for the experience, for which I have been very grateful ever since. Thank you for the chance to once again meet up with old mates and hear such humorous stories that I will remember for ever.'

From the start of the project it was clear that Ngamatea had been a formative influence on all who had spent time there and that the ties forged in their youth had endured over the years. They described those who ran the property, who had survived under all sorts of hardships and who had kept people in work in tough times, as 'the backbone of the country'. And they were more than that to their young musterers, as Gay Withers explained. 'They look back almost with love – yes, with love – on the Roberts. It was tough, but they were family. They are still close to Jack, and devastated about Margaret. If anyone had done anything to Margaret back then they'd have given their lives for her.' And as Dave Withers put it, 'You work on other properties, get into responsible situations – but I've never ever lost that understanding of caring and working with people – all those attributes from working at Ngamatea, working in those extreme conditions.'

The loss of the big leasehold blocks out the back, the inevitable split of Ngamatea and Timahanga, the development of the grasslands and the replacement of horses with motorbikes, meant 'some of the challenge, the mystique, whatever you like to call it of Ngamatea as a whole started to disappear. It just wasn't the same.' Part of the enjoyment of living up there was being a part of that country.

Many talked of the aura and the magic of Ngamatea. It was a self-contained world, the experience of a lifetime. It was tough and it was a challenge, especially for young musterers, but, as Dave Wedd made clear, they were content with their lot. 'All that tussock, and pigs, and deer, and wild sheep. Yes, and wild dogs. Amazing country. I don't think there's anything like it anywhere else. Ngamatea has got the bush – the atmosphere. Atmosphere, that was what it was all about. You go there, and it holds you; you never forget it. You never forget Ngamatea.'

Notes

ONE Owhaoko

1. Interviews with R. A. L. Batley at Moawhango, 29 March 1993 and 1 Nov 2003. Tony Batley researched and wrote extensively on the land and the people of his area.
2. *New Zealand Government Gazette* (*Gazette*), 1 June 1875, p. 385; MS-Papers-2928 (Folders Collection), Alexander Turnbull Library (ATL).
3. Judgments of the Court, 10 Dec 1885, *Appendices to the Journals of the Legislative Council* (*AJLC*), 1887, sess II, No.1, p. 2. The validity of the memorial of ownership was questioned by Robert Stout, Memorandum on Owhaoko and Kaimanawa Native Lands, 18 May 1886, *Appendices to the Journals of the House of Representatives* (*AJHR*), 1886, G9, pp. 4–6.
4. New Zealand Archaeological Association Site Record Form, N123/6.
5. Owhaoko (80,790 acres) and Owhaoko No. 1 (17,160 acres) from Renata Kawepo, Owhaoko 1b (2,860 acres) and Owhaoko B (13,465 acres) from Noa Huke, and Owhaoko A (40,395 acres) from Ihakara te Raro, Karaitiana te Rango and Retimana te Rango; Owhaoko block records, Native Land Court (NLC).
6. E. J. Studholme, *Coldstream: The Story of a Sheep Station on the Canterbury Plains 1854–1934*, Christchurch, 1985; 'Family History' from notes by W. P. Studholme, March 1933, written by Sir Henry Studholme, Wembury House, Plymouth, typescript, Dec 1973, Studholme Family Papers, Coldstream; interviews with Joe Studholme at Ngamatea, 15 Jan 2004 and Coldstream, 2 and 3 Mar 2004.
7. *Gazette*, 30 Jan 1879, p. 165.
8. Studholme, *Coldstream*, pp. 11, 19.
9. *Hawke's Bay Almanack*, 1883, p. 90.
10. R. T. Warren Correspondence 1888–1903, MS-Papers-0272-11, ATL; Warren letters, Studholme Family Papers, Coldstream.
11. See e.g. Alan Ward, *A Show of Justice: Racial 'Amalgamation' in Nineteenth Century New Zealand*, Auckland, 1973; David V. Williams, '*Te Kooti Tango Whenua*': *The Native Land Court 1864–1909*, Wellington, 1999.
12. Judgments of the Court, 10 Dec 1885, *AJLC*, 1887, sess II, No.1, pp. 2–3; Owhaoko Judgment 10 Oct 1888, John Studholme Papers, MS-Papers-0272-18, ATL.
13. Stout memo, *AJHR*, 1886, G9, pp. 1–23; Report, evidence etc. of parliamentary select committee into Owhaoko and Kaimanawa Native Lands, 13 Aug 1886, *AJHR*, 1886, I8; Buller's statement, *AJHR*, 1887, Sess I, G1.
14. Judgment – Owhaoko 1 July 1887, *AJLC*, 1887, sess II, No.1 pp. 3–7; Owhaoko Judgment 10 Oct 1888, *Napier Evening News* 11 Oct 1888, in MS-Papers-0272-18, ATL.
15. *AJLC*, 1887, sess II, No.1, pp. 6, 7.
16. Owhaoko Judgment, 10 Oct 1888, MS-Papers-0272-18, ATL; see also Renata's statement 23 July 1886, *AJHR*, 1886, I8, pp. 83–4.
17. *AJLC*, 1887, sess II, No.1, p. 3.
18. Notebooks, MS-Papers-0272-38 and -39, ATL.
19. Studholme, *Coldstream*, p. 23.
20. S. W. Grant, 'Airini Donnelly, 1854/55?–1909, Ngati Kahungunu woman of mana, landowner', in *The Dictionary of New Zealand Biography, Volume Two 1870–1900*, Wellington, 1993, p. 121.
21. Joe Studholme at Ngamatea, 15 Jan 2004.
22. Notebook, MS-Papers-0272-38, ATL.
23. MS-Papers-0272-05 and 0272-26, ATL.
24. MLC-WG acc W1645 3/1910/217, Archives New Zealand (ANZ).
25. Studholme, *Coldstream*, pp. 12–13, 24, 69; Studholme 'Family History'.
26. *Gazette*, 21 Nov 1907, pp. 3387–8; Owhaoko Block, MS-Papers-0272-01, ATL; Block Records, D5, pts 2, 3, 4; D6, pt 2; D7, NLC.
27. *Gazette*, No. 63, 1914, p. 2664.
28. Conversation with Marion Donnelly at Hastings, 14 June 2004; correspondence between Beatrice Findlay (*née* Shaw)

29 Block Records, D5, pts 2, 3, 4; D6, pt 2; D7, NLC; *Gazette*, 10 Jan 1924, p. 101, 23 Jan 1931, p. 123; various memos 1930/31, Owhaoko Gift Blocks – 1916–1975, AAMX w3430 26/1/12 pt 1, ANZ.
30 Interview with John and Margaret Ruddenklau at Bay View, 15 Dec 2003.
31 Conversations with and photos from Bob and Ian Watherston, 26 Feb and 5 Apr 2004; memo, 6 Nov 1928, AAMX w3430 26/1/12 pt 1, ANZ.
32 *Gazette*, 23 Jan 1931, p. 123.
33 Memos, 10 Nov 1910 and 23 June 1911, MLC-WG acc w1645 3/1910/217 and /246, ANZ.
34 Native Land Amendment and Native Land Claims Adjustment Act 1930, s 25(1); *Gazette*, 10 Jan 1918.
35 Sec. M. Affairs to Sec. Tuwharetoa Trust Board, 16 Jan 1957, AAMX w3430 26/1/12 pt 1, ANZ.
36 Memo, 20 Sept 1971, ibid.
37 Native Land Amendment and Native Land Claims Adjustment Act 1930, s 25(1); ibid.
38 See Patrick Parsons, 'Te Ahiko, Raniera, ?–1894, Ngati Te Upokoiri and Ngati Kahungunu historian', in *The Dictionary of New Zealand Biography, Volume Two*, pp. 513–14.
39 Napier Minute Book No 36, 12 Dec 1894, p348, NLC.
40 Spear, p. 14.
41 Interview with Jack Roberts at Timahanga, 28 Oct 2003.
42 Tape of Johnny Boyd and Jack Roberts, n.d; conversations with Ernie Kirkman, 19 May 2004 and Phyllis Walker, 20 May 2004 and 23 July 2004; Spear, p. 20.
43 Owhaoko C3, MLC-WG w1645 3/4537, ANZ.
44 Owhaoko vol 3, undated letter, possibly Sept 1933, MA Wang acc w2140 wh605 box 39, ANZ.

TWO Fernie Brothers and Roberts

1 Miriam Macgregor, *Mangaohane: The Story of a Sheep Station*, Hastings, 1978, pp. 43–4.
2 It is said the blocks were transferred to him by the owners of Mangaohane in settlement of a debt. He probably took them over in 1929, although the title transfer is dated 15 Aug 1933.
3 Interviews with John Roberts at Wanganui, 7 May 2003; and Noel and Eileen Roberts at Dannevirke, 7 May 2003.
4 John Roberts; Jack Roberts.
5 *Gazette*, 29 Oct 1931, p. 3035.
6 Interview with Jack Roberts (jnr) at Hastings 5, 6 and 7 Aug 2003.
7 Jack Roberts.
8 The wages book shows J. F. Roberts running Ngamatia (the old spelling) from 9 March 1931.
9 Spear, p. 23.
10 Interview with Lou and Joan Campbell at Taihape, 9 Sept 2003.
11 Interview with Gordon and Eunice Mattson at New Plymouth, 27 and 28 March 2003.
12 Jack Roberts.
13 Noel Roberts.
14 John Roberts.
15 Lou Campbell.
16 Interview with Ray White at Whirinaki, 12 March 2004.
17 Noel Roberts.
18 The name is also spelled Johnson and Johnston in the wages register.
19 Winnie Roberts, 'Notes', mss, n.d.
20 Interview with Gordon Maxwell at Ngamatea, 29 Oct 2003. Gordon's father worked with George Everett on Te Awaiti.
21 Gordon Mattson.
22 Winnie Roberts, 'Notes'.
23 Noel Roberts.
24 D. A. Bathgate, *Yesterday (Inanahi)*, Hastings, Hart Printing, 1970, pp. 110–13.
25 Lou Campbell.
26 John Roberts.
27 Winnie Roberts, *Spectrum* documentary, National Radio, 28 Feb 1976.
28 Gordon Mattson.
29 Lou and Joan Campbell.
30 Lou Campbell.
31 Winnie Roberts, 'Notes'.

THREE Lawrence and Winnie

1 John Roberts; Jack Roberts.

2. Interview with D'Arcy Fernandez at Pukawa, 2 and 3 Nov 2003.
3. Winnie Roberts, *Spectrum* documentary.
4. Gordon Mattson.
5. Ray White; Jack Roberts.
6. Ray White.
7. *Brassica oleracea*, a type of kale.
8. Gordon Mattson.
9. Interview with Bill and Evelyn Wells at Hastings, 4 Aug 2003.
10. Winnie Roberts, 'Notes', and *Spectrum* documentary.
11. Jack Roberts.
12. Eileen [Joll] Roberts.
13. Eunice [Joll] Mattson.
14. Jack Roberts.
15. Ray White.
16. Jack Roberts.
17. Winnie Roberts, *Spectrum* documentary.
18. D'Arcy Fernandez.
19. Winnie Roberts, *Spectrum* documentary.
20. Winnie Roberts, 'Notes', and *Spectrum* documentary.
21. D'Arcy Fernandez.
22. Interview with Gordon Grant at Hatuma, 5 March 2004.
23. Gordon Grant.
24. Jack Roberts.
25. D'Arcy Fernandez.
26. John Roberts.
27. Jack Roberts.
28. Owhaoko Gift Blocks – 1916–1975, w3430 26/1/12 pt 1, ANZ.
29. Ibid.
30. Ibid. The estimate of the area grazed is very conservative.
31. Interview with Ray Birdsall at Whakaroa, Taupo, 16 April 2003.
32. Jack Roberts.
33. Interview with Bill Cummings at Taihape, 8 Sept 2003.
34. D'Arcy Fernandez, Jack Roberts.
35. Interview with Bruce Atchison at Wanganui, 6 May 2003.
36. Interview with Peter van Dongen at Ngahinapouri, 24 June 2003.
37. Winnie Roberts, *Spectrum* documentary.
38. In 1960 I stayed with the Wickershams at Newport Beach in California and benefited from their visit to Ngamatea. They repaid to me the hospitality they had received from Winnie and Lawrence.
39. Interview with Dave and Gay Withers at Tuahu station, Ruakituri Valley, 30 July 2003.
40. Interview with Frank Brady at Olrig station, 8 May 2003.
41. Interview with Paul Thomas at Taradale, 4 Aug 2003.
42. Dave Withers.

FOUR **The Muster**
1. Bill Cummings.
2. Interview with Don and Alison Hammond at Hastings, 8 May 2003.
3. Interview with Arthur McRae at Matawhero, 31 July 2003.
4. Bruce Atchison.
5. Bruce Atchison.
6. Interview with Dave Wedd at Hastings, 6 Aug 2003.
7. Interview with Dixie McCarthy at Taihape, 6 Sept 2003.
8. Interview with Trevor Topliss at Taihape, 6 Sept 2003.
9. Don Hammond.
10. Bill Cummings.
11. Interview with Don McLean at Palmerston North, 17 Oct 2003.
12. Interview with Ivan and Elaine Chapman at Hastings, 9 May 2003.
13. Ray Birdsall.
14. Ivan Chapman.
15. Ray Birdsall.
16. Ivan Chapman.
17. Interview with Alex Lindsay at Waipawa, 2 Aug 2003.
18. Ivan Chapman.
19. Bruce Atchison.
20. Frank Brady.
21. Don Hammond.
22. Dave Withers.
23. Bruce Atchison.
24. Bill Cummings.
25. Jack Roberts and Dave Wedd.
26. Alex Lindsay.
27. Dave Wedd.
28. Jack Roberts; Dave Withers.
29. Bruce Atchison.
30. Ivan Chapman.
31. Frank Brady.
32. Ivan Chapman.
33. Bruce Atchison.
34. Barry Crump, at the start of his deer culling career, stayed in the Manson hut and wrote about it in *The Life and Times of a Good Keen Man*. His version of Bruce's handiwork (p. 38) differed

from Bruce's; for my money, Bruce got it right.
35 Ivan Chapman.
36 Paul Thomas.
37 Alex Lindsay.
38 Dave Withers.
39 Dave Withers.
40 Ivan Chapman.
41 Dave Wedd.
42 Ray Birdsall.
43 Frank Brady.
44 Dave Withers.
45 Frank Brady.
46 Arthur McRae.
47 Bill Cummings.
48 Bruce Atchison.
49 Bruce Atchison.
50 Jack Roberts.
51 Ray Birdsall.
52 Jack Roberts.
53 Ray Birdsall.
54 Interview with Craig Hammond at Hastings, 8 May 2003.

FIVE Incidents and Accidents
1 Gordon Mattson.
2 Winnie Roberts, *Spectrum* documentary.
3 Interview with Maxie Birdsall at Whakaroa, Taupo, 23 June 2003.
4 Winnie Roberts, *Spectrum* documentary.
5 Dave Withers; Dave Wedd.
6 Ray Birdsall.
7 Ray Birdsall.
8 Jack Roberts.
9 Dave Withers.
10 Interview with Alan Bond at Matawhero, 31 July 2003.
11 John Roberts.
12 Interview with Joe Larrington at Waipawa, 1 Aug 2003.
13 Jack Roberts.
14 Dave Withers.
15 Ray Birdsall.
16 Conversation with Park Pittar at Ngamatea Reunion, 23 July 2004.
17 Peter van Dongen.
18 Ray Birdsall.
19 Ivan Chapman; Frank Brady.
20 Dave Withers.
21 Interview with Richard Whittington at Pukehou, 10 March 2004.
22 Dave Withers.
23 Winnie Roberts, 'Notes'.
24 Noel Roberts.
25 Bill Cummings.
26 Dave Withers.
27 Paul Thomas.
28 Ivan Chapman.
29 Dave Wedd.
30 Frank Brady.
31 Dave Wedd.
32 Ivan and Elaine Chapman.

SIX Jack and Timahanga
1 Interview with Jack and Jenny Roberts at Timahanga, 28 Oct 2003.
2 Bruce Atchison.
3 Alan Bond; Jack Roberts; Dave Wedd.
4 Jack Roberts.
5 Jenny Roberts.
6 Jack Roberts.
7 Jack Roberts.
8 Jack and Jenny Roberts.
9 Jack Roberts.
10 Jenny Roberts.
11 Interview with Lance Kennett at Whareroa station, West Taupo, 11 April 2003.
12 Interview with Bob and Bev Ralph at Blackhead Road, Wallingford, 21 July 2004.
13 Dave Wedd.
14 Jenny Roberts.
15 Jack Roberts.
16 Jenny Roberts.
17 Occupational Safety and Health.
18 Jack Roberts.
19 Brodifacoum, a second-generation anti-coagulant.

SEVEN The Development Years
1 Interview with Larry Cummings at Taihape, 8 Sept 2003.
2 Maxie Birdsall.
3 Ivan Chapman.
4 Conversation with Denis Goulding at Ngamatea Reunion, 23 July 2004.
5 Conversation with John Bennett at Ngamatea Reunion, 23 July 2004.
6 Maori Land Court (MLC) Block Records.
7 Larry Cummings.
8 Conversation with Noel Merwood at Taihape, 8 Sept 2003.
9 Phone conversations with Rex Tobeck, Palmerston North and Kevin Schimanski, Wanganui.
10 Ray Birdsall.
11 Interview with Eric Brooking at Waimarama, 10 March 2004.

12. Peter van Dongen.
13. Interview with Chris and Des Page at River Road, Taupo, 5 September 2004.
14. Minister of Maori Affairs to Minister of Lands, 18 March 1968, AAMX w3430 26/1/12 pt 1, ANZ.
15. Bob Tutaki to Registrar Ikaroa Land Board, 6 Nov 1934, MLC-WG w1645 3/3214, ANZ.
16. Lands and Survey memo, 2 Dec 1969, AAMX w3430 26/1/12 pt 1, ANZ.
17. Secretary to Minister of Maori Affairs, 17 May 1973, ibid.
18. *Dominion*, 11 June 1973.
19. Ray Birdsall.
20. Interview with Terry Apatu at Taradale, 6 Oct 2004.
21. Interview with Elaine Mahoney at Forest Park, Taupo, 16 Feb 2004.
22. Terry Apatu.
23. Elaine Mahoney.
24. Elaine Mahoney.
25. Des Page.
26. Super-giant discs: seven a side, blades about a foot apart and 30 inches high; with the cut right on they sink into the ground 8 or 10 inches and chop up tussock, scrub and other vegetation. Off-set discs: 12 feet wide, 24-inch blades 8 or 9 inches apart.
27. A big cultivator, more a levelling tool.
28. Richard Whittington.
29. Conversation with Paul Hughes at Ngamatea, 29 Oct 2003.
30. Richard Whittington.
31. Interview with Sheila and Gordon Maxwell at Ngamatea, 29 Oct 2003.
32. Gordon Maxwell.
33. Elaine Mahoney.
34. Elaine Mahoney.
35. D'Arcy Fernandez.
36. Elaine Mahoney, D'Arcy Fernandez, Richard Whittington.
37. D'Arcy Fernandez.
38. Gordon Maxwell.
39. Terry Apatu.
40. Richard Whittington.

EIGHT Cooks, Cops and other Characters

1. Ivan Chapman.
2. Bruce Atchison.
3. Frank Brady.
4. D'Arcy Fernandez.
5. Dave Wedd.
6. Ray Birdsall.
7. Jack Roberts.
8. Peter van Dongen. Mountain oysters: lamb testicles.
9. Paul Thomas.
10. Jack Roberts.
11. Bill Cummings.
12. Gordon Grant.
13. Ray Birdsall.
14. Ray Birdsall.
15. Interview with Lindsay McRae at Bay View, 15 Dec 2003.
16. Winnie Roberts, 'Notes' and *Spectrum* Documentary.
17. Ibid.
18. D. A. Bathgate, *Hawke's Bay Herald-Tribune*, 17 June 1967, p. 6.
19. Jack Roberts.
20. Lester Masters, *Scent of Manuka Smoke and other Poems*, Hart Printing, Hastings, n.d.
21. D'Arcy Fernandez.
22. Dave Wedd.
23. Paul Thomas.
24. Dave Withers.
25. Don Hammond; Bruce Atchison; Don McLean.
26. Jack Roberts.
27. Don McLean. Ian Sinclair wrote about this episode in *Boot in the Stirrup*, an account of his time in the high country. In his version he won the encounter with Gunner.
28. Dave Withers.
29. Conversation with Jack Bindon at Ngamatea Reunion, 23 July 2004.
30. Conversation with Harry Bimler at Ngamatea Reunion, 23 July 2004.
31. Interview with Sid Drinkrow and George Appleby at Napier, 28 and 29 April 2001.
32. George Appleby.
33. Jack Roberts.
34. Larry Cummings.
35. Des Page.
36. Craig Hammond.
37. Ray Birdsall.
37. Ray Birdsall.
39. See Vernon Wright, 'Wild Horses', *New Zealand Geographic* 1, Jan–March 1989, p. 63.

NINE Margaret

1. Joe Larrington.
2. Jack Roberts.
3. Alex Lindsay.
4. D'Arcy Fernandez.
5. Conversations with Belinda Dakin at

Tai Tapu, Canterbury, 1 March 2004, and by phone, 22 Sept 2004.
6. Interview with Kerry Chalmers at Otamauri, 20 July 2004.
7. Belinda Dakin.
8. Kerry Chalmers.
9. Belinda and Steve Dakin in Ngamatea House Visitors' Book, 1 Sept 1995.
10. Bruce Atchison.
11. Kerry Chalmers
12. Alan Bond.
13. Eleanor unveiled the plaque 'placed in recognition of a substantial contribution given the YMCA by Mrs Fernie and her husband, Walter, who have been supporters of the organisation for many years'. *Wanganui Herald*, 13 June 1974, p. 7.
14. Terry Apatu.
15. Terry Apatu.
16. Paul Thomas.
17. Interview with Wendy Jacobs at Norsewood, 16 Dec 2003.
18. Wendy Jacobs.
19. Interview with Graham and Margaret Lunt at Waipukurau, 29 Feb 2004.
20. Gordon and Sheila Maxwell.
21. Oral Decision and Sentencing of Judge G. A. Fraser in the District Court Held in Taihape, 5 July 2000, CRN: 0067003096/7.
22. Graham Lunt. An entry in the Ngamatea Visitors' Book 4 April 1992 recorded 'SAS exercise, 85 from Linton, 4 teams of tracker police dogs'; and 6 April 1992 'army helicopters arrive and fugitives released, a day late due to bad weather – heavy snow.'
23. Interview with Ren Apatu at Hatuma, 23 July and 5 Oct 2004.
24. Ren and Sally Apatu.
25. Interview with Steve Kelleher at Ngamatea, 26 Oct 2003.
26. Steve Kelleher.
27. Conversations with Paul Hughes, Oct, Nov 2004.
28. Steve Kelleher.
29. Larry Cummings.
30. Conversations with Martin Brenstrum at Ngamatea, 13 and 14 March 2004.
31. Renata Apatu.
32. Steve Kelleher.
33. Correspondence between Beatrice Findlay and Marion Donnelly, Christmas 1995 and 14 July 2000.
34. Alan Bond.
35. Jack Roberts.
36. Arthur McRae.
37. Gordon and Sheila Maxwell.
38. Wendy Jacobs.

Bibliography

Unpublished
Correspondence between Beatrice Findlay (*née* Shaw) and Pat and Marion Donnelly.
Correspondence with Heather Woollacott (*née* Findlay), Parktown, South Africa.
Folders Collection, MS-Papers-2928, Alexander Turnbull Library.
Maori Affairs Department – Wanganui Archives, Archives New Zealand.
Maori Land Court – Wanganui Archives, Archives New Zealand.
Napier Minute Book No. 36, 1894, Native Land Court.
New Zealand Archaeological Association Site Record Form, N123/6.
Ngamatea Visitors' Book.
Ngamatea Wages Registers.
Oral Decision and Sentencing of Judge G. A. Fraser in the District Court Held in Taihape, 5 July 2000.
Owhaoko Block Records, Native Land Court.
Owhaoko Gift Blocks, Land Corp Accession, Lands and Survey Archives, Archives New Zealand.
Roberts, Winnie, 'Ngamatea', *Spectrum* documentary, Replay Radio, National Radio, 28 Feb 1976.
Roberts, Winnie, 'Notes', n.d.
Spear, John, 'Memories of Inland Patea', typescript, n.d.
Studholme Family Papers, Coldstream.
Studholme, John, Papers, MS-Papers-0272, Alexander Turnbull Library.
White, Ray, 'One Man in His Time: The Memoirs of Ray White', manuscript, n.d.

Official
Appendices to the Journals of the House of Representatives.
Appendices to the Journals of the Legislative Council.
New Zealand Government Gazette.
New Zealand Statutes.

Books and Articles
Annabel, Ross and Marion Donald, *The Heart of the Parapara and Field's Track*, Wanganui, Parapara-Field's Track Historical Society Inc, 2002.
Bathgate, D. A., *Yesterday (Inanahi)*, Hastings, Hart Printing, 1970.
Caselberg, John (ed.), *Maori is my Name: Historical Maori Writings in Translation*, Dunedin, John McIndoe, 1975.
Crossley, Jeanette, *Matt's Last Muster: A Shepherd's Journey*, Palmerston North, Ekrub, 1992.
Fernandez, Natalie, *Tussock Fever*, Auckland, Boughtwood Printing, 1973.
Grace, John Te H., *Tuwharetoa: The History of the Maori People of the Taupo District*, Wellington, A. H. & A. W. Reed [1959], 1966.
Grant, S. W., 'Donnelly, Airini, 1854/55?–1909, Ngati Kahungunu woman of mana, landowner', in *The Dictionary of New Zealand Biography, Volume Two 1870–1900*, Wellington, Bridget Williams Books/Department of Internal Affairs, 1993, p. 121.
Hawke's Bay Almanack.
Hogan, Helen (ed.), *Renata's Journey: Ko te Haerenga o Renata*, Christchurch, Canterbury University Press, 1994.
Holden, Philip, *Station Country: Back-Country Life in New Zealand*, Auckland, Hodder & Stoughton, 1973.
Laurenson, S. G., *Rangitikei: The Day of Striding Out*, Palmerston North, Dunmore Press, 1979.
Lethbridge, Christopher, *Sunrise on the Hills*, Auckland, Hodder & Stoughton, 1971.
Macgregor, Miriam, *Mangaohane: The Story of a Sheep Station*, Hastings, Herald Tribune

Print, 1978.
Masters, Lester, *Tales of the Mails*, Hastings, Hart Print, 1959.
Masters, Lester, *Scent of Manuka Smoke and other Poems*, n.d.
Newton, Peter, *Big Country of the North Island*, Wellington, A. H. & A. W. Reed [1969], 1973.
Newton, Peter, *Big Country of the West and North*, Wellington, A. H. & A. W. Reed, 1972.
Parsons, Patrick and Angela Ballara, 'Kawepo, Renata Tama-ki-Hikurangi, ?–1888, Ngati Te Upokoiri and Ngati Kahungunu leader, missionary', in *The Dictionary of New Zealand Biography*, *Volume One 1769–1869*, Wellington, Allen & Unwin/Department of Internal Affairs, 1990, pp. 218–19.
Parsons, Patrick, 'Te Ahiko, Raniera, ?-1894, Ngati Te Upokoiri and Ngati Kahungunu historian', in *The Dictionary of New Zealand Biography*, *Volume Two 1870–1900*, Wellington, Bridget Williams Books/Department of Internal Affairs, 1993, pp. 513–14.
Rangitikei Heritage Trail Working Party, 'Inland Patea, Taihape – Hastings, Rangitikei', Heritage Trails, 2004.
Robinson, Tony (ed.), *West to the Annie: Renata Kawepo's Hawke's Bay Legacy*, Waipukurau, CHB Print, 2002.
Sinclair, Ian, *Boot in the Stirrup*, Wellington, A. H. & A. W. Reed, 1973.
Studholme, E. J., *Coldstream: The Story of a Sheep Station on the Canterbury Plains 1854–1934*, Christchurch, 1985.
Tait, G. A. (ed.), *Farms and Stations of New Zealand*, vol. I, Wn-73, Auckland, Cranwell, 1957.
Ward, Alan, *A Show of Justice: Racial 'Amalgamation' in Nineteenth Century New Zealand*, Auckland, Auckland University Press, 1973.
Wheeler, Colin, *Historic Sheep Stations of the North Island*, Wellington, A. H. & A. W. Reed, 1973.
Willems, Hans, *North Island Back Country Huts*, Auckland, Halcyon Press, 1998.
Williams, David V., *'Te Kooti Tango Whenua'*: *The Native Land Court 1864–1909*, Wellington, Huia Publishers, 1999.
Wright, Vernon, 'Wild Horses', *New Zealand Geographic* 1, Jan–March, 1989, pp. 53–67.

Newspapers
Dominion.
Hawke's Bay Herald-Tribune.
Napier Evening News.
Wanganui Herald.

Index

Photographs are indicated in bold

accidents: with horses, 70, 135, 137; planes, 132–5; vehicles, 136, 138, 259, 263
accountants, 45, 71, 165
acreage, 3, 9, 15, 18, 23, 34, 82, 165, 166, 210
aeroplanes, 45, 132–4, 143, 166, 168, 185, 233, 240
Aga stove, 62, 181
Air Force, 88, 199, 245
Aldridge, Jack, 143, 187
Andrews, Dave, 182
Andrews, Norm, 192
Annie, the, *see* Gentle Annie
Apatu, Kate, 195
Apatu, Margaret, *see* Roberts, Margaret
Apatu, Nathan, 195
Apatu, Renata, 195, 260, 263, 274, 284; and Ngamatea, 264–5, 266, 267, 272; education, 264, 267; marriage, 277
Apatu, Sally (Baines), 277
Apatu, Terry, 161, 184, 195, 196, 248, 260
Apatu, Wilson and Molly, 161
Appleby, George (and Nita), 233, **234**, 235–6, 279
Army/SAS, 32, 55, 56, 136, 143, 154–5, 204, 205, 206, 240, 263–4
Arnett, Alice, 30
Atchison, Bruce, 89, 91, 97, 100, 101, 106, **107**, 110, **111**, 113, 114, 120, 148, 164, 211, 220, 246, 278, 281
Atkins, Ralph, 93, 94, 266

back country, 18, 23, 24, 51, 60, 81, 88, **98**, **99**, 102, 112, 114, **115**, **116**, 122–5 *passim*, 182, 184
Baker, Tudor N., 20
Barnes, Seth (and Rose), 33, 34, 49, 247
Barrett, Sonny, 92, 231
Bates, 'Chips', 33, 34–35, 40
Bathgate, Dr D. A., 45, 109, 225–6, 275
bathing/hygiene, 108, **109**, 110, 113
Batley, Tony, 9, 13, 14, 86, 280
Baylis, Bob, 49
Beamish, Joy, **275**
Begorra (pony), **244**
Belt, Ken, 78
Bennett, John, 183, 193
Bibbenluke Stud, **79**, 80
Bimler, Harry, **111**, 233
Bindon, Jack, **111**, 233
Birch, Azim and William, 8, 13
Birdsall, Anthony, 179, 181, 194
Birdsall, Maxie (Mackenzie), 129–31, 135, 154, 179, 194, 261, 275; at Ngamatea, 179–82 *passim*, 238

Birdsall, Ray, 5, 101, **111**, 122, 124, 129–31, 137, 141, 157, 275, 281, 285; and cops, 155, 238–41; and hunting, 100, 178–80; and poachers, 237–41; and SAS, 154–5; farm at Taupo, 194; head-musterer, 82, 99, 144, 213; manager at Ngamatea, 133–4, 136, 165, 179, 182, 185, 188, 237–41 *passim*, 253; musterer, 97, 104, 114, 135, 149, 220; stock manager at Haupiri, 179
Birdsall, Tim, 181
birthdays, 21sts, 149–51
Blowfly Gully, 52
blue duck, 95
Bond, Alan, 122, 140, 164, 192, 227, 248, 276, 282
Boomer, the, 114
Booze (dog), 48, **49**
Bower, Dave, **111**, 214, **216**
bowser, 238, 239
Boyd, Arthur, 25–28, 33
Boyd family, 24–25
'Boyd' homestead, 25–26, 27, 28
Boyd hut, **107**, 153–4
Boyd, Jessie (Kirkman), 25, 26
Boyd, Johnny, 25–27, 281
Boyd Rock, 2, 285
Boyd's Bush, *see* Kaimoko
boys, the, 88, 94–95, 106, **107**, 110, 119, 130, 140–1, 146, 147–8, 151–2, 164, 213–4, 255, 284; and police, 151, 152; and Taihape, 151–3, *see also* musterers
Brady, Frank, 101, 102–3, 106, **107**, 108, 114, 152, 153, 255, 266, 282
bread, 26, 213, 230
Brenstrum, Martin, 270, 271
Brenton-Rule, John, 185
bridges: Pohokura, 167, 236; swing, 77, 150, 166, **167**, 220; Woolwash, 88, 89, 206
Brittan, Guy, **107**, 120
Brooking, Eric, 187, 199, 282–3
Brooking, Hazel, 187, 188, 194, 282
Brooks, Keith, 136
Broughton, Wi, 17
Brown, Phil, 51, 52
Bryant and May, 108, 124
Bulled, John, 134
bullocks, 12, **21**, 25
Burke, Matt, 172, 264
Burnell, Noel, 49

'Cactus', *see* Gillan, Reg
calf marking, 131
Cameron, E. H., 11
Cameron, Peter, 83, 84, 122, 211
camp oven, 91, 104, 143
Campbell, Joan, 51, 280

Campbell, Joe, 136, 236
Campbell, Lou, 34, 44, 49, 51, 52, 230, 280
Carmichael, A. B., 20
Cascade beer/brewery, 94, 98
cats, 139–40, 148, 246; feral, 271
cattle, 52, 166, 199, 217, **250**, 261, 269–70
CB radio, 144, 199, 206
chaff, 40, **90**, **93**, 230, 231
Chalmers, Kerry, *see* McRae, Kerry
Chapman, Elaine, 156–7, 284
Chapman, Ivan (Shorty), 96–97, 100, 101, 128, 155–7, 211, 226, 227, 229, 243
Chase, Mary, 150
Chase, Tommy, **147**
Chatham Islands, 70, 98, 200–01
Cheesman, Hilton, 92
Chesterhope, 30
Christison, Andrew, 271
Christmas, 78, 119, 130, 184, 188, 215
Churchill, 29
circus, 152–3
Clarke, Alan and Carolyn, 170
Clifford, Dave, 127
Cobber (packhorse), 101, 231–2
Coldstream, *see* Studholme family
Coleman, Garry, 183
Colenso, William, 8
Collins, Woody, 142
Comet, the, 33, 174
conservation, 81, 190–1, 210, 261
cooks, 97, 118–9, 164, 183, 211–7, 226
Cooksley, Al, 150
Cornwall, Jack, 92, 93, 96, 99, 110, 221
Correspondence School, 46, 65, 66, 181
court cases, 175, 239, 241, 262
Coutts, Tony, 146
Crafar, Marty, 196, 261
Craven family, 32, 68
Craven, Nanny, 66
Craven, Ted and Joyce, 62, 83, 151, 165
Craven, Winnie, *see* Roberts, Winnie
Cribb, Colin, 191, 192
crops/cropping, 40, 61, 184, 260, 268, 274
Crowe, Jim, 127–8
Crown, the, 18, 22–4, 25, 80–82, 123, 174, 182, 189–91
Crownthorpe (Matapiro) reunion, 285
crutching, 125–6
Cummings, Bill, 93, 94, 124, 149, 184, 218, 226, 280
Cummings, Larry, 184, 185, 237–8, 270, 271

dagging, 119, 246
Dakin, Belinda, *see* Wilson, Belinda
Dampney, Hartrey, 93
deer, 55, 73–4, 98, 100, 145, 233, 234–5, 238, **262**; live capture, 237, 240; sikas, 113, 176; skins, 57, 73, 113, 176, 179
deer cullers, 55, 151, 232–3

Depression years/Slump, 41, 50, 51
dip/dipping, 121–2, 185
discs/discing, 74, 75, 76, 198, **204**
docking, 119
dog kennels, **103**
dogs, 61, 93, 94, 100, 114, 115–6, 118, 126, 139–41, 184, 228
Dolman, Lou, 237
Donnelly, Airini, 15, 17
Donnelly, G. P., 12, 17, 18, 19, 29
Donnelly, Marion, 275, 283
Donnelly, the, *see* Makorako
Doole, Bernie, 49, 56
Dorreen, Neil, 83
D'Ott, Dorothy, *see* Roberts, Dorothy
D'Ott family, 33
double- or triple-deckers, 55–56, **59**, 228, *see also* hermits, roughies
Dowden, the, *see* Mt Dowding
Dowding, C. H., 11
Doy, Stanley, 51
Drinkrow, Sid (and Lee), 233, **234**, 235–6, 279
drovers/droving, 121, 149, 282
Duffell, George, **122**
Duggan, Peter, 41
Duley, Edwin, 202
Dumbleton, Jack, 212
Dunn, Barry, 134, 182
Dunstan, Ashley, 60, 186–7
Dunstan, Ron, 60
Durand, Gerry, 45

earthquake, Hawke's Bay, 34, 54
Edgecumbe, Oliver, 258, 259
Ellison, Henry (surveyor and mapmaker), 14
English, Ted, 245
Erewhon, 8
Everett, George, 42, 49, 50, 69, 200, **221**; and his dogs, **139–40**; character, 43, 139, 140, 212
ewes, 5, 61, 79, 83, 119, 120, 122, 124, **125**, 183, 220, 268, 274
extreme conditions, 123, 288
family tree, Fernie–Roberts, 31
farriers, 186
fences, fencing, 123, 153–4, 166, 170, 182, 186, 187, 199, 206, 208, 269
Fernandez, D'Arcy, 69–71, 72, 95, 141, 181, 203, 206, 227, 251, 254, 282; and hunting, 73–74
Fernandez, Natalie (McLeod), 74, 97, 181, 203, 282
Fernie, Annie (Mrs Roberts), 30, 32
Fernie brand, **147**, 148
Fernie Brothers and Roberts, 81
Fernie, David, 30, 32, 234
Fernie, Drummond, 1, 30, 33, 34, 38–39, 40, 44, 53, 54, 77, 234

Fernie, Eleanor (Hallas) 30, 44, 165, 248–9
Fernie family, 29–30, 149
Fernie, Helen (Forsyth), 30, 44
Fernie, Joan, 30
Fernie, John, 29–30
Fernie, Walter, 1, 30, 44, 54, 58, 78, 82, 161, 165, 234, 248–9
ferrets, 176–7, 270, 271
Findlay, Beatrice (Shaw), 19–20, 275, 283
Findlay, Christian (and Lizzie), 275
fire, 21, 40, 118, 124, 239; Ohinewairua fire, 204, 205, 206, 207, 208–9
firewood, 12, 62, 127, **129**, 232
First Aid Post, 174
fish/fishing, *see* trout
Fisher, Bruce, 188
fog, 114, 146
Forest Service, 24, 124, 190, 204, 205
Forks, Spur, 123, 144
Forrester, Anne and Chris, 187
Fraser, Ken, 128–9, 155–6
Fraser, Pet, 129, 156–7
freehold: Ngamatea, 3, 81, 82, 190; Pohokura and Te Koau, 33, 76, 165; Timahanga, 168–9, 174
frost, 131–2, 179
funerals, 128, 164, 217, 251, 277–8, 284

Gardner, W. R., 20
Gayton, Denis, 185
general hands, 55, **56**, 119, 180, 186
Gentle Annie, 2, 136, 173
Gift Block, 3, 23–24, 80–82, 190–1
Gilbert, Ted, 94
Gillan, Reg ('Cactus'), 128, 222
Gillett, Colin, 191, 192
gold, 8
Gold Creek (Panoko), 8, 113, 153, 285
Golden Hills, 8, 20, 26, 60, 90, 91, 116, **117**, 125, 153, 206, 231,
Golden Hills hut, 42, **93**, 101, 105, **106**, 143–4, 183, 227, **284**, 285, **286**
Gold Knob, 182
Goulding, Dennis, 183
Grant, Beryl, 75–76, 219
Grant, Gordon, 74–77 *passim*, **167**, 219, 221, 281
Green, Peter, 195
Gregory-Hunt, Boe, 246
Gretna Hotel, 21, 50, 119, 152
Gunner (packhorse), 232
guns/rifles, 25, 73, 144–5, 154, 175, 213–4, 238–40, 246

hacks, *see* horses
Hamilton, Hauparoa, 28
Hammond, Alison (Dorreen), 92
Hammond, Craig, 126, 188, 283
Hammond, Don, 92, 153–4, 217–8, **221**, 283

hāngi, 121, 187, 254, **255**
Harker, Old Tom, 116, 217–**9**, 220–1
Harker, Young Tom, 92, 94, 116, 217, 218–9, 231
Harkness, the, 112, 124, 153–4, 182
Hatch, Mrs (cook), 65, 66–67
Hawke's Bay Show, 78, 115, 237
Hawkins', 261
Hawkins' hut, 184
hay, 166, 186, 208, **209**, 270
Hazeldine, Les, 185
helicopters, 133–4, 154, 155, 205–6, 237, 240, 241, 261
Henaghan, Kerry, 188, 189, 196, 206
hermits, 51, **59**, *see also* roughies, double-deckers
Hildebrant, Pauline, 244
Hill, Peter, 215
hillbilly, 104, 114
Hindus, 33, 40, 53
Hiwiopapakai (Trig F), 13
Hogget block, 132, 133–4, 182, 206, 227–8
hoggets, 119, 122, 126, 264, 269
Horotea (lake), 206
horse trucks, 258–9, 263, 278
horses/hacks, 88–91, 116, 117, 128, 135, 197, 213, 220–1, **250**; breaking, 94, 221–2; breeding, 258–59, 266
Horsfeld, Albert, 276
Hughes, Jim (cook), 212
Hughes, Paul, 198–9, 267, 268, 274
Huke, Noa, 8, 9, 15
huts, 95, 105–6, *see also* individual huts
hydro, 116, 191, **192**, **193**, 243

Ikawetea Stream, 7, 33
Inland Patea, 7, 8, 9, 13, 14, 18, 24, 57
Iona College, 243, 224–5
Isaacson, Henry, 56
Isaacson, Ted, 49, 51, 52, 57

Jacobi, Miss, 38
Jacobs, Wendy and Robyn, 257–60, 265–6, 277
Jeffery/Jeffrey, Albert (Jeff), 54–55, 222–3, 224
Jimmy's, 34, 74, 76, **171**
Jimmy's Creek (Timahanga Stream), 33, 163, 217, 226
Johnston/Johnstone, Jerry, 41–42, 53, 55
Joll, Eileen, *see* Roberts, Eileen
Joll, Eunice, *see* Mattson, Eunice
Jones, Bill, 121, 228, 254, **256**, 286

Kahungunu, 7
Kaimanawa block, 91, 116
Kaimanawa Face, 145, 182
Kaimanawa Forest Park, 123, 190
Kaimanawa hut, 95, 123, 144, 231

INDEX 299

Kaimanawa–Oruamatua block, 8, 13
Kaimanawa Range, 1, 9, 15
Kaimanawa stream (Waingakia), 95
Kaimoko (Boyd's bush), 12, 33, 166, 167, 174, 190
Kate (packhorse), 230
Kaweka Forest Park, 124
Kaweka Range, 1
Kawepo, Renata, 8, 15, 16, 17, 24; and John Studholme, 16; and Land Court, 9, 14, 15; and Lindauer portrait, 16, 284; and surveys, 13
Kelleher, Keryl, 266
Kelleher, Steve, 5, 266–9 *passim*, 274, 277
Keller, Geoff, 236
Kelly block, 195, 266–7
Kennett, Alan, 98, 111, 155, 186
Kennett, Lance, 4, 141, 155, 170–1, 188, 189, **283**, 284–5
King Billy (pony), **64**
Kirkpatrick, Colin, 54, 72, 211, 212
Koroneff, Nick, 240
Kuripapango, 8, 14, 75

Lager, Jack, 50
Laidlaw, Rob, **283**, 285
Lake block, 82, 166
lambing beats, 89, 101, 129, 153
lambs/lambing, 121, 268–9; fat, 5, 183, 185
lamps/lighting, 62–3
land development loans, 168, 194, 209
land tenure, *see* freehold, leasehold
Lands and Survey Department, 24, 81, 174
Larrington, Joe, 138, 160, 237, 242–3
Larrington, Rita, 242
Law, Peter, 186
leasehold, 76, 80
leases: Owhaoko, 8, 10, 14, 17, 18, 20, 22; Mangaohane, 16, 18; Ngamatea, 34; Timahanga, 25, 165
Lee, Arthur and Glenda, 261, 264
Leonard, Jimmy, 24, 25, 33, 153
Lethbridge, Chris, 95–6
Lewis, Mihi-Mere, 247
Lindsay, Alex, 100, 125
Log Cabin, 56–57, 106–7, **108**, 136, 146, 230
London, Tom and Pat, 75, 76
Londontown, 76
Looker, Peter, 122, 136, 143, 161, 276, 287
Lumsden, Dave, 135–6
Luckie, F. D., 18
Lunt, Graham, 260–4 *passim*, 266
Lunt, Margaret, 260, 261
Lyall family, 64–65, 66, 159
Lyall, Hugh (Otupae), 26
Lysaght, Andy, 209

Macdonald, Rosie, 234, 245
Mahoney, Elaine (Shannon), 195–6, 202, 205, 254
Mahoney, Phil, 170, 195–7, 198–9, 206, 253, 260, 264
mail, 173, 272
mailbox, 166, 173, 272
Makorako (the Donnelly), 9, 93, 104, 131–2, **286**
Maney, R. D., 8, 9
Mangamaire Stream, 15, 93, 123
Mangamingi hut, 113
Mangamingi Stream, 93, **99**, 104
Mangaohane, 5, 8, 16, 17, 18, 26, 29, 51, 55, 189, 199
Mangatapiri, 32, 182, 234
Mangataramea, 12, 24
Manson, the, 60, 101, **102**, 111–2, 123, 124, 140–1, 182
Manson Creek, 60, 101, 111
Manson hut, 20, 110, **111**
Maori Affairs Department/Minister, 24, 81, 191
marijuana, 240
Masters, Lester, **225**, 226–7
Matahuki, Wi and Cookie, 170
Matapiro, 102, 199
Matapiro (Crownthorpe) reunion, 285
Matches, Guy, 172
Mattson, Eunice (Joll), **64**–66, 275
Mattson, Gordon, 4, 34, 42–46 *passim*, 49, **64**, 65, 66, 69, 127–8, 139, 234, 237, 275, 279, **280**; foreman, 55, 61
Maxwell, Gordon, 199–200, 201–2, 261, 262, 266, 272, **283**, 285
Maxwell, Sheila, 199–200, 254, 261, 262, 272–3, **283**, 285
Mays, Olive, 74
McCarthy, Dixie, 91, 93, 177–8, 281
McDermott, Dan, 188
McGregor, Scotty, 55
McLean, Don, 94–95, 232
McLellan, Jack, 134, 183
McLellan, Peter, 183, 185
McLeod, Nat, *see* Fernandez, Natalie
McMahon, A. (Gus, Mac), 69, 137–8
McPherson, Murray, 197
McQueen, Dr Ross, 9
McRae, Arthur, 89, 117–8, 121, 161, 227–8, **276**, 277, 282
McRae, Kerry (Mrs Chalmers), 244, 245, 277
McRae, Lindsay, 221–2, 281
McSweeney, Leo, 150, 214–6, 217, 246
Merwood, Noel, 186
Mills (carriers), 69, 72
Minto, Forbie, 89
Miracle (hack), **64**
Missy, 4, 64, 67–68, 216, 243, 247, 248, 278, 279, 283, *see also* Roberts, Margaret
Moatapuwaekura, 151
Moawhango people, 7, 13, 14

Moeangiangi, 30
Moody, Adrian, 271
Moore, Ian, 94, 104–5
Morrin, Tom, 29
Mossman, Max, 287
motorbikes: 2-wheel, 237–8, 264; 4-wheel, 266, 271, 283, 285
Mott, Morrie, 99, 100, 101–2, 110, 145, 231, 243, 275
Mt Cameron, 11, 56, 58, 59
Mt Donnelly, *see* Makorako
Mt Dowding (the Dowden), 11, 93, 104, 230, **286**
Mt Michael, 12, 52, 93, 115, 123
Mounganui, 183, 184, 249, 258
Mudguts (roping horse), 221
Munro, Jock, 55
Murimutu, 11, 17
Murphy, Alf, 49
muster, the, 2, 21, 51, 60, 88, 100, 103–4, 114, 116, 118, 161; crutching, 125–6; shearing: Timahanga, 220; straggle, 56–9, 124–5, 182, 183
musterers, 88, 91, **96**, 99, 103, 107–8, 111, 116, 118, 128, 140, 143–4, 161, 177, 230, 250; head musterer, 88, 94
mustering: at Ngamatea compared with South Island, 92; of fine-wools compared with Romneys, 92, 118; season, 56, 88, 116, 125, 131

Napier Boys' High School, 159–60
Native Land Court, 9, 13, 14, 17, 18, 25
newspaper, 55, 109–10, 173, 181
Ngamatea, 26, 34, 38, 40, 51, 82, 126, 130, 171, 179, 210, 237, 247, **287**; atmosphere of, 256, 286, 288; back boundary post, 261, 283; buildings, 34, 35, 187, 195, 196, 202, 273; deer farm, 240, 270; land development, 183–4, 185–6, 197–9, 200, 201–02, 204, 210; magic of, 2, 251, 286; vastness, 46, 69–70, 197
'Ngamatea family', 2, 5, 279–84
Ngamatea football team, 49–50
Ngamatea homestead: new, 180, 181, 195; old, 12, 34, 35–**36**, 62, 180, 276
Ngamatea House, 252, 253, 254; and John Scott, 251–2; warming, 254
Ngamatea stables, 35, 186, 196, 197, 232
Ngamatea store, 69, 71, 272
Ngamatea Swamp, 189, 191, 210
Ngamatea woolshed: new, 121, 202, 254–7; old, 34, 35, 82–83, 120, 202
Ngamatea workshop, 35, 196, 197
Ngamatia, 12, 18, 20, 50, 79
Ngaruroro River, 15, 104, 112, 113
Ngati Hotu, 7
Ngati Kurapoto, 7, 15
Ngati Maruahine, 15

Ngati Tama, 7, 15
Ngati Te Upokoiri, 14, 15
Ngati Tuwharetoa, 7, 8, 190–1
Ngati Whiti, 7, 8, 15
Ngati Whititama, 15
No Man's Land, 153
'No Sale' (poem), 223–4
Norwood, Annie (Mrs Fernie), 29
Noxious Weeds Act, 23
nurses, 75, 129, 172, 261; district, 137, 188

oats, 40, 182
Occupational Safety and Health Act (OSH), 175
O'Connor, Brian, 94
Ohinewairua, 8, 204, 264
Olrig, 153, 266, 282
Omahu people, 13, 14
Oppatt family, 61
Oracle (pony), **64**, 244
Orange, Rex, 234, 235–6
Otamoa, 30, 33, 149, 185, 249
Otupae block, 119, 179, 187, 240
Otupae Range, 29, 46–47, 166
Otupae Station, 5, 8, 29, 33, 55, 64, 184, 189
Owhaoko and Kaimanawa-Oruamatua Reinvestigation of Title Act 1886, 14
Owhaoko block: and leases, 8, 10, 14, 15–16, 17, 18, 20, 22; and the Crown, 18, 22–24, 80–82, 123, 182, 189–91; Maori association with, 15, 22–23, 27; physical characteristics; 9, 23–24, 81, 189; subdivisions, 8, 14, 15, 17, 80–81, 184, 189–90
Owhaoko Station, 16, 17, 79, 189
Owhaoko trig, 255, 286

pack-boxes, 143, 229, 230, 232
packhorses, 40, 88, **89**, 90, 91, 98, 102, 124, 136, 182, 225, 227, **229**–32
packman-cook, 88, 90, 103, 104, 142–4
pack-tracks, 33, 81, 136, 285
paddock names, 77, 187
Paerau, Horima, 8, 9
Page, Chris and Des, 187–8, 194, 197, 238
Page, Rebecca, 188
Panoko Stream, *see* Gold Creek
Paramahao (Pohokura bush), 33, 167
Parkinson, Atholl, 56
Parsons, Wayne, 186
Passmore, Bob, 150
Pastoral Run No.2, 3, 25, 169
Paton, John and Norah, 161, 164–5
pest control, 176, 270–1
Peter's, 41, 82, 182, 233, 285
Peter's hut, 145
Pine, Te Hiraka, 23
Pinnacles, the, 141, 256, **263**
Pittar, Park, 141, 161

INDEX 301

poachers/poaching, 175–6, 237–8, 239–41
Pohokura, 7, 24, 33, 53, 75–6, 119, 122, 150–1, 166, 167, 235, 236; cookhouse/whare, 38, 213, 215, **216**; outstation of: Mangaohane, 29, 213; Ngamatea, 38, 213, 217, 220; Timahanga, 165, 213–4; woolshed, **170**
Pohokura bush, *see* Paramahao
Pohokura hot spring, 38, 172
Pohokura lake, 171, **177**; water skiing, 171
police, 49, 127, 128, 133–4, 173, 174, 175, 216; and Ngamatea boys, 151, 152; and poachers/poaching, 237–41 *passim*, 260, 262, 263; and Ray Birdsall, 238–41
possums, 176, 180–1, 270, 271
Potham, Terry, 93, 104–05
Poverty Creek, 123, 261
power, mains, 194
Preston, Roger, 236
Pukeroa, 30, 149

quarries, 199, 201
QEII Trust, 167, 190
Quilter, Harold, 69, 237, 286
Quirke, Jack, 50

Rabbit Act, 23
rabbit-proof fence, 26, 117, 154
rabbit rates, 4, 24, 189
rabbiters, 41, 58, 60, 155, 176–7, 184; huts, 183, 184
rabbits, 18, 20, 21, 22, 139, 184, 208
Ralph, Bev and Bob, 25, 172, 281
rams, 1, 38, **78**, 80, 83; fat lamb sires, 183
Rangitikei River, 15, 104, 123, 141, 142, 204, 208, 238
reading matter, 107–08
Redford, Gavin, 183
reunions, 254–7, 285–7
Riddiford, Ivor, 93
rifles, *see* guns
Right Hand Fork (of Ngaruroro), 113
river/stream crossings, 112–3
roadmen, 61, 63
roads: Napier/Hastings–Taihape, 2, 21, 34, 136, 173, 238, 258, 286; station, 61, 97, 190, 241, 247–8; Timahanga–Pohokura, 77, 167, 170, 174, 178, 225–6
Roberts, Alan, 169, **170**, 172
Roberts, Barry and David, 45, 68, 77, 159
Roberts, Dorothy (D'Ott), 32–33, 34, 36, 45, 53, 68
Roberts, Eileen (Joll), 64–67 *passim*, 281
Roberts Ellen (Abraham), 30
Roberts, Jack (J.F. snr), 30, 32, 52, 230; at Ngamatea, 34, 36, 37, 51; at Timahanga, 33–34; death, 53
Roberts, Jack (J.R.), 4, 64, **68**, 134, 136, 149–50, 152, **177**, 182, **225**, 239, 276, 278, 281, 287; at Ngamatea, 102, 106, 119, 160, 228; at Timahanga, 161, 165–9, 171, 172, 174, 177–8, 190, 213; early years, 53, 63–**64**, 67–69, 237; education, 64–65, 66, 70–71, **72**, 159–60, 161; in South Island, 160; marriage, 161
Roberts, Jackie (Wood), 172
Roberts, Jenny (Paton), 162–3, **177**, 178; and cooks, 164, 169–70, 172; and education, 161; and family, 169; and first aid, 174; and travellers, 173–4; at Timahanga, 163–4, 173; marriage, 161
Roberts, John (head shepherd), 196, 264, 283, 285
Roberts, John/Johnny (J.F. jnr), 33, **35**, **37**, **46**, 47–48, **49**, 50, 75, 77–78, 92, 138, 149
Roberts, Johnny, 164, 169, 172
Roberts, Joseph, 30, 32
Roberts, Karyn (Hughes), 172
Roberts, Lawrence, 3, 52, 53, 81, **84**, **86**, 96, 119, 126, 179, 216, 222; and Jack, 64, 159, 160; and Margaret, 64, 245, 248; and the boys, 82–83, 86–87, 94–95, 120–1, 131, **147**, 150, 154, 164, 287; and the muster, 60, 88, 123, 160–1; and wool, 57, 67, **79**, 83; army service, 32, 86; at Mangatapiri, 32, 53, 234; birth, 30, 32; character, 60, 61, 76, 85, 86, 94, 122, 124, 136–7, 160, 211; illness and death, 161, 164; marriage, 53; sense of humour, 58, 91, 101, 108, 139, 148–9, 150
Roberts, Margaret/Missy (Mrs Apatu), 4, 53, **68**, 179, 194, 195, 196, 216, 248, 274, 275, **276**, 277, 284, 286; and horses, 242, 243–4, 257–9, 266, 276, 277; and hunting, 257, 258; and Ngamatea, 165, 251–6, 258, 260, 263–4, 267–8, 271; and poachers, 259–60; and staff, 259, 261, 266–7, 268, 273, 284; and wool, 2, 183, 246–7, 248, 267; at Hatuma, 257–8, 265–6; early years, **64**, 67–69, 70–71, 237, 242–6; education, 66, 70–71, 243, 244–5, 246–7; illness and death, 4, 277–8; marriage, 161
Roberts, Noel, 33, 34, **35**, 37–8, 40, **49**, 50, 61, 65, 74, **75**, 191, 281
Roberts, Peter, 169, **170**
Roberts, William, 30, 32
Roberts, Winnie (Craven), 2, 3, 4, 41, 45–**46**, 48–49, 64–65, 69, **79**, **84**, **170**, 196, **249**; and accidents, 84, 127–8, 131–2, 135; and horses, 46–47, 50; and Mangatapiri, 53; and Ngamatea, 46, 54, 63, 67–68, 71–73, 105, 131–2, 148, 222–3, **225**, 250; and Te Awanga, 164–5, 249–50; and the boys, 84, 135, 137; and visitors, 84–85; and work, 47–48, 61, 62; birth, 32; character, 83–84; illness and death, 250–1; marriage, 53; sense of humour, 42, 53, 66, 229
Robertson, Ali, 188, 238

Robin, Ike, 22, **77**, 78, 200
Robinson, Happy, 137–8
Rock camp, 60, 112, 230
Rosie (hack), 64
roughies, 57,**59**, 60, 78, see also double-deckers, hermits
Royce (packhorse), 226
Ruahine Forest Park, 175
Ruahine Range, 15, 25, 33
Ruanui, 13, 17
Ruapehu, 104
Ruddenklau family, 20–21, 34
Ruddenklau, John, 20–21, 281
Ruhanui (Jimmy Leonard's), 24
Rusthall, 30, 149

Sandilands, Howard, 144–5
SAS, see Army
Sayers, Kevin, 136, 161
Schimanski, Kevin, 186
schoolroom: at Ngamatea, 70, **72**; at Pohokura, 161, **162**
Scott, Barney, 52
Scott, John: and house at Hatuma, 265; and Ngamatea House, 251–2, **253**, 254
scrub, 166; burning, 40, 171; crushing/discing, 74–76; cutting, 33, 38, 40, 167
Search and Rescue, 132, 237, 282
Selwyn, Rex, 261
Sharp, Dave, 271
Shaw Annie (Mitchell), 18, **19**, 20
Shaw, Basil, 19
Shaw, Beatrice, see Findlay, Beatrice
Shaw family, 18
Shaw, Guy, 18, **19**, 20
shearing, 69, 78, 119, 120, 197; straggle shear, 125
shearing contractors, 77, 78, 121
Sheehan, Denis, 236
sheep: Booroola, 185, 196; composite, **269**; Corriedale, 161, 171, 177, 197; English Leicester, 79–80, 119, 183; halfbred, 1, 17, 26, 51, 80, 112, 123, 159, 183, 267–8; merino, 1, 17, 51, 79–80, 112, 119, 123, 183, 197; Romney, 196, 264; studs, 79–80, 120, 183; turned out for winter, 122–3
sheepfold, 9
shelter, 34, 36, 61, 67, 260
shepherding: compared with mustering, 119; fine-wools compared with Romneys, 264
shepherds, 88, 150, 185, 186, 213, 217, 261, see also musterers
silage, 186–270
Sinclair, Ian, 94, 232
Sinclair, Ken, 80
Singh, Daljeet, 270
Sisam, Pete, 206
Slump, the, see Depression years
Smedley, 150, 260, 264, 266

Smith, John, 161
Smith, S. P. (surveyor), 13
snow, 48, 56, 94, 97, 123, **125**–6, 140, 146, 156,182, 186, **254**, 259, 260, 268, 276
Sommerton, Bill, 161, 227–8
South Island: mustering compared with North Island, 92
Sparrowhawk Range, 33
Spear, John, 19, 24, 26, 34
Spiral, the, 76, 77, 136–7, 167, 216, 236
Springvale, 114
staff, 271–2; attachment to Ngamatea, 2, 54, 69, 77, 105, 178, 179, 194, 233, 272, 273, 276–7, 279, 281, 285, 286, 288; attitudes, 103, 160–1, 272; casuals, 191, 269; lost, 144, 145–6
station books, 45, 55, 71, 181
Steed, Eric, 276
Steele, Jack, 136
Stevens, George, 49, 50
stock numbers, 50–51, 79, 166, 183, 185, 274
stores, 71–73, 261, 272
Stout, Robert (premier and attorney general), 14
Strip, the, 2, 61
Studholme brothers, 10–11
Studholme family, 3, 10, 12, 13, 16, 17, 79; and Coldstream, 10, 11, 16, 17, 284; and Murimutu, 11, 17; and Owhaoko, 10–13, 15–17; and Ruanui, 13, 17; and Waimate, 10, 11
Studholme (Owhaoko/Mangataramea): homestead ('Boyd's'), 11, 12, 24, **27**; woolshed, **12**, 35, 202
Studholme, Joe, 4, 17, 275, **282**, 283–4
Studholme John, 10, 11, 15–16, 17, 18
Studholme, Sue, 4, 283–4
Sturm, Dick, 97
Sunny Face, 210
survey dues, 18
surveys, 13
Surveyor's Rock, see Rock camp
Swamp block, 60, **96**, 117, 133, 134, 206, 210

Tamakopiri, 7
Tamateapokaiwhenua, 7
Taruarau hill, 2, 126, 173, 238
Taruarau River, 7, 52, 57, 126, 132, 134, **167**, 220, **229**
Taruarau valley, 60, 117
Tarzan (packhorse), 142, 232
Tauwheketewhango/Tawake Tohunga, see Peter's
Taylor, Murray, 136, 161, 286
Taylor, Rev. Richard, 8
Taylor, Stan, 110
Te Awaiti, 43, 200

Te Heuheu Tukino IV, Horonuku, 15
Te Heuheu Tukino V, Tureiti, 23
Te Koau, 25, 29, 33, 75, 165
Te Momo, Nani, 99
Te Naonao, Paramena, 15
Te Oke, Hira, 8, 14
Te Rango, Karaitiana, 8, 15
Te Rango, Retimana, 8, 15
Te Raro, Ihakara, 8, 15
Te Uamairangi, 15
Te Wanikau, Anaru, 15, 17, 24
telephone, 186; party-line, 63, 164; station, 63, 163
television, 156
Three Musketeers, 233–6
thieves, 238
Thomas, Paul, 85, 131, 146–7, 150–1, 214–15, 255–7, 283
Tikitiki Bush, 9, 13, 40, 82, 127, 206, 236, 239, 241, 276, 285
Tikitiki hut, 85, 139, 140, **156**, 235
Timahanga block, 24, 33; and Pastoral Lease, 33; and the Crown, 25, 174–5
Timahanga outstation, 38, 70, 126
Timahanga Station, 14, 33, 163, 165–6, **168**, 169, 171, 178, 184; buildings, 33, 34, 161–2, **163**, **165**, **169**, 170; deer farm, 176; development, 166–8; woolshed, 33, 149
Tin Kettle Spur, 93, **116**, 123
Tit, the (Te Toka a Ruawhakatina), **180**, 280, 282, 285
Tobeck, Rex, 186
topdressing, 166, 168, 185–6, 198
Topia, Kingi, 23
Topliss, Trevor, 92, 281
Tornado (hack), 237
trace clipping, 52, 57
tractors: 183, 184, 241, **265**; crawlers, 39–40, **41**, 75, 76, 138, 183, 221; John Deere, 146, 184, 197, 198, **204**
transport: Lumsden, 94; Mills, 69, 72; Overland, 98, 164
trespass, 175, 239
Trooper (packhorse), 230–1
trout, 141–2
tuberculosis (TB), 176, 270, 271
Turkey George, 114
tussock, 2, 9, 34, 46, 81, 123, 180, 250, 251, 256, 270, 273, 278; types of, 51, 92, 124, 184, 189, 246
Tuwhakaperei, 7
Tuwharetoa Trust Board, 81

Utu (film), 203–4

van Dongen, Joy, 178, 282

van Dongen, Peter, 142–4, 178, 282
venison, 73, 179–80, 237, 241
Vincent, Margot, 80, 246
visitors, 84–86, 275

wagons, 40, **129**, 273
Waiouru, 11, 149
war: First World War, 32, 86; Second World War, 54, 55
Warren, R. T., 11, 12, 60
washing, 108
Watherston, Bob and Ian, 21
Watherston, Jock, 21, **22**, **23**, 77
Watt, Ashley, 92, 94, 95, 98, 104, 111, 114, 153, 218, 227
Watts, Peter and Jenny, 266
Wedd, Dave, 89, 95, **99**, 107, 126, 132, 145, 151–2, 161, 227–8, 276, 281, 288
weddings, 53, 161, 163, 200, 248
Wells, Bill and Evelyn, 61–62, 177, 218
Wellwood family, 67, 83, 159, 245
wetas, 57, 95, 107
Whakamarumaru, *see* Mt Michael
Whimp, Tom, 36–37, 40, 41, 54, 196, 202
White, Dr, 134
White, Ray, 61, 136, 275, 280, 286–7; and rabbiting, 58; and straggle muster, 56–59; musterer, 55–56, 60
White's block/valley, 82, 137, 184, 237, **262**
Whitikaupeka, 7, 8, 14, 15
Whittington, Richard, 197–8, 199, 201, 206, 209–10, 273
Wickersham, Harold and Mary-Lou, 85
Wild Dog Spur, 261
wild dogs, 12, 114–5, 134
wild pigs, 25, 231, 236
wild sheep, 23, 51, 52, 58, 59
Williams, H. B. (Otupae), 29
Wilmott, Tom, 127, 128
Wilson, Belinda (Mrs Dakin), 244–6
Wilson, Tau, 43, 77, 78
Winnie (packhorse), 229, 230
Withers, Dave, 107, 108, 126, 131, 132, 137, 145–6, 147, 150, 226, 232, 281–2, 288
Withers, Gay, 288
Wong, Harry, 73
wool, 185, 196, 269
Woollacott, Heather, 283
Woollaston, John, 186
wool trucks: early, **121**; late, 278
woolwash: at Ngamatea, 191; at Tikitiki, 9
Woolwash Stream, 191, **192**
Wright family (N.S.W.), 80
Wright, Gordon, 107, **147**

Zigzag, the, 93, 123